AF553903

ENVIRONMENT AUDIT

ENVIRONMENT AUDIT

Prof. A.K. Shrivastava
Chairman-cum-Chief Executive "CTREE"

IN ASSOCIATION WITH
COUNCIL FOR TRAINING & RESEARCH IN ECOLOGY & ENVIRONMENT
C-18-19, Qutab Institutional Area, New Delhi-1100016

A P H PUBLISHING CORPORATION
5, ANSARI ROAD, DARYA GANJ
NEW DELHI-110 002

Supported by : All India council for Technical Education (AICTE) Under the Ministrry of Human Resource Development, Government of India.

Published by
S.B. Nangia
A.P.H. Publishing Corporation
4435–36/7, Ansari Road, Darya Ganj,
New Delhi-110002
Phone: 011–23274050
e-mail: aphbooks@gmail.com

2026

Typeset by
Ideal Publishing Solutions
C-90, J.D. Cambridge School,
West Vinod Nagar, Delhi-110092

Printed at
DIVINE DIGITAL PRINTERS
Ansari road Daryaganj Delhi-110002

Dedicated to

My Mother Radhika Devi &
Father Late Dr. G.N. Sahay

PREFACE

It was 1992, when the Government of India announced the policy abatement of pollution and, recognised "environment auditing" as one of the instruments for achieving the objectives of integral environmental considerations into decision making at all levels. The statement was "Industrial concerns and local bodies should feel this they have a responsibility for abatement of pollution. It was also made clear further that the procedure of an environment statement will be introduced in local bodies, statutory authorities and the public limited company to evaluate the effect of their policies, operations and activities on the environment, particularly compliance with standards, and the generation and recycling of wastes.

But the past one decade has shown a least progress in relation of environmental auditing of the local authorities and even of polluting industries which was made mandatory for them to conduct auditing and submit the auditing report periodically. The reason, what I understand, is the lack of political will and inadequacy of fund, specially in terms of local authority's environment audit. The commitment of the development countries to assist developing countries and countries faces economic transition may be the one another reason.

The fact is however, very discouraging this since last one decade we could not create an awareness for environment auditing among the corporate sector, local authorities and the pressure group-the people. We could not prepare and organise external auditing team or consultant environment auditor or trainer to train the authorities responsible for environment auditing.

Now, the decision makers or policy planners should not talk about "sustainable development" on papers. This is the high time, when we must be sincere about the sustainable development measures and formulate the strategy to preserve the elements of environmental stock to meet the essential needs of the day and tomorrows as well. And, it is needless to repeat that it urges regular monitoring of environmental scenario of the country.

In fact, the specific objective of this presentation is to help guide the industrial and local authority's personnel-responsible for environmental auditing and also to prepare environmental auditors for the future need.

We have no way to escape, if we are not having our concern with the environment as per the norms-it may be a temporary relief which may cause an unpleasant reality in future.

With a great hope, we look forward your right step.

Environmentally yours'

A.K. Shrivastava
Chairman : CTREE.

ACKNOWLEDGEMENTS

The undersigned would like to thanks to the **All India Council for Technical Education** (AICTE), MoHRD, Govt. of India for their encouraging support in getting this book published. The associate publisher also be thanked for the same, who have shown courage and enthusiasm to publish the book.

The undersigned thanks are due to a number of UK based environmental auditors whose expertise has contributed in this book, either through their reports, articles or books on different aspects of the audit, especially in the field of environmental auditing in local areas.

The author (undersigned) is thankful to CPCB for providing important data/case studies in relation to environment audit, which helped the undersigned in including industrial audit procedure in India.

The undersigned also record his gratitude to the CTEE staff and well-wishers.

A.K. Shrivastava
CTREE.

ENVIRONMENT AUDITING: A FRAME OF EDUCATIONAL/TRAINING PROGRAMME

Over recent years, it has become obvious to both industry and the general public alike that the damage being caused to all aspects of the global environment is unsustainable. The old approach to pollution prevention of control "react and cure" has been recognised as an unsophisticated way of dealing with environment problems, and is being superseded by a new "anticipated and prevent" approach, which is for more cost effective. This realisation has led to the development of a wide range of environment management techniques specifically aimed at reducing industry's impact on the environment and to demonstrate to the public that they are doing all they can reduce this to a minimum.

Increasing numbers of organisations, in both the public and private sectors, are now beginning to adopt environment and to demonstrate to the public that they are doing all they can to, reduce this to a minimum.

Specialists from a wide range of different fields will require an understanding of Environment Audit procedures to ensure that they are undertaken in the most effective manner. Ecologists, Engineers, Soil Scientists & Economists must use their technical expertise to undertake Environment Audits; Lawyers must establish and operate a legal basis for the procedures, industrialists must formulate their own environment policies to take account of the need for Environment Audits. It is important that full realisation of the potential of Environment Auditing to minimise and prevent industrial pollution impact of the environment is not hindered by a lack of trained Environment Audit specialists.

The rapid development of Environment Audit over the years has led to a gap in the market for trained Environment Auditors. This gap will be exacerbated with the introduction of the Bureau of Indian Standard on Environment Management. Indian Environment Management and Audit scheme and the International

Environment Management and Audit Scheme, and the International Environment Management Standard, all of which rely heavily on trained environment auditors.

Environment Auditing : Introduction

Environment Auditing is required for facilitating management control over environment practices and for assessing compliance with companies' policies, which would meet regulatory requirements. International Chamber of Commerce defined it as a management tool comprising a systematic, documented, purist and objective evaluation, who shows how well environment organisations - management and equipment are performing as safeguard of the environment.

Full Detail of Course

ENVIRONMENT AUDITING

Module 1 : The Nature of Environment Audit.

The term Environment audit means a range of different things to different people. The aims of this introductory module are to define Environment Auditing and to provide participants with some idea of the history behind Environment Auditing and why it developed. This module also provides an introduction to environment science, so that participant can identify the environment problems posed by human activities and can, therefore, better understand the need for environment management and environment auditing.

Module 2 : Types of Environment Audit

There are ranges of different types of procedure, which fall under the term environment audit, from energy audits to environment management audits and from product audits to local authority audits. This module will identify these various types, define what they involve and provide some idea of who uses them.

Module 3 : Environment Management Systems

Environment Audit is not the sole answer for reducing a company's impact on the environment. It forms part of an over environment management systems which is designed to ensure that all aspects of a company's operations which have, or could have, an impact on the environment, are reduced to a minimum. This module will provide details of what is involved in environment management system, how to set them up and how to ensure that they operate correctly.

This module will also provide participants with an understanding of the environment consultancy sector and some of the activities that all environment consultants are expected to undertake in their day to day operations. These would include time keeping and the costing of projects.

Module 4 : Environment Audit Procedure

Although the general procedure for undertaking Environment Audit is fairly well known, it is quite difficult to obtain details of systems that organisations carrying out audits actually use. These organisation, particularly environment consultancies often treat their methodologies as commercially sensitive, given the investment or resource necessary to develop them. The aim of this module is to provide participants with the information required to undertake Environment Audits. The basic Environment Audit procedure consists of three phases; **pre-audit, site visit and post-audit**. The module will provide details of what is involved in each of these phases. Questionnaires or protocols are valuable, tools used by Environment Auditors to obtain information concerning the facility being audited. This module will also provide information on how to produce questionnaires, what type of questions to ask, and what kind of ensures to look out for.

Module 5 : Environment Economics and Risk Assessment

The use of an environment economic approach to decision making has been recommended to resolve issues that arise during environment audits. This module is designed to provide participants with a basic understanding of the concepts involved in environment economics, risk Assessment forms an essential clement of any decision making process. Consequently, a grounding in this important field if also provided.

The Module will provide information on :

Environment economics; the economic problems; scarcity and choice, the allocation of resources in a free market; demand, supply and the process mechanism. Divergence of private and social calculations; private costs and benefits *Vs* social costs and benefits. The concept of externalities and the absence of a market therein. Pollution and environment destruction, attempts to estimate their economic costs. Pollution controls; physical regulation *Vs* Economic incentives via taxes and subsidies. Cost-benefits analysis, an outline and assessment of the technique its strengths and weaknesses. Multiplier impact analysis; the multiplier concept and its application to projects at the local/regional level.

Risk Assessment

Types of risk; expected value; degree of risk; risk preference; technical risk; stochastic assessment of atestrophic events, scientific assessment of risk.

Module 6 : Established Environment Management Procedures.

Environment Auditing is now being adopted by many other countries. For example, certain of the former Eastern & Western countries which include India are now adopting legislation which requires Environment Audits essentially to be undertaken for certain types of industry. In India, a system or voluntary Environment Auditing has been introduced which becomes mandatory for companies, which endanger the public. This module will provide participants with an idea of which countries are adopting a more formal approach to Environment Auditing thus enabling them to offer their skills as auditors to a wider market do.

Thus module will provide detailed know how of "eco-audit" scheme.

Modules 7 : The Environment Management Context

Environment Auditing is but one of a number of procedures that are now available to organisations wishing to reduce the impact they have on environment. This module will focus on two of these procedures EIA, Risk Analysis and Assessment. This will ensure that participants understand the concept of the procedures.

Module 8 : Case Studies in Environment Management

The aim of this module is to provide participants with examples of the environment management and Environment Audit systems and the problems, which may be faced by companies in implementing them. By simulating the environment problems posed by companies and examining what steps can be used to avoid or negate them, participants learn to use the skills they have been taught in earlier modules. We shall focus on the collection and assessment of monitoring date, which must occur if a company is to demonstrate that its environment management system is operating correctly.

SUMMARY

The Executive/National Diploma components of the course is of one year duration (although this can be extended) commencing in June leading principally to the award of a Diploma of the CTREE successful completion. There may be an opportunity for some participants to proceed to higher degree.

COURSE STRUCTURE

Course

Module 1	The Nature of Environment Audit
Module 2	Types of Environment Audit
Module 3	Environment Management Systems
Module 4	Environment Audit Procedure

Optional

Module 5	Environment Economics and Risk Assessment
Module 6	Established Environment Management Procedures
Module 7	The Environment Management Context
Module 8	Cast Studies in Environment Safety Management

(A National/Executive Diploma programme in the stream of Environment Auditing and Environment Impact Assessment being conducted by the Council for Training & Research in Ecology and Environment, New Delhi through Distance Learning method)

CONTENTS

Preface vii

Acknowledgements ix

Part-One : Environment Auditing in Local Areas

- Sustainable Development 1

1. Environment Auditing 7
2. Frame Work of the Environmental Auditing 17
3. Elements of Environmental Stock 35
4. Nature Conservation 45
5. Energy 63
6. Transport 77
7. Land Use Planning 89
8. Conservation Audit 93
9. Pollution Control 97
10. Waste and Recycling 115
11. Eco-consumerism–The Purchase Audit 125
12. Community Awareness 143

Part-Two : Analysis of Current Practice

13. Developing an Auditing Strategy 165
14. Management of the EA Process 195
15. Future Prospect of Environment Auditing 223

Part-Three : Environment Audit in India

16. Environment Auditing in India 233

17. Environment Audit Procedure 247

18. Environment Auditing: What, Why and How 255

19. Environment Auding : An Overview 261

20. Auditing Programme in Major Polluting Industries 271

Case Studies 287

Annexure 317

Glossary 333

Appendices 337

Index 361

Part-One

Environment Auditing in Local Areas

SUSTAINABLE DEVELOPMENT

The holistic approach towards contemporaneous environmentalism is "sustainable development" - this was the religious wish of the group of environmentalists during the "Earth Summit in Rio, 1992" where it was discussed in a greater way. The theory was finally accepted that the earth has limited capacity to carry population growth, urbanisation and its pressure, technical innovation - industrialisation and thus adverse effect of pollution on the Air, Water and Land.

In the face of this, governments have started taking initiatives to do more and not only the environmental window dressing just for the sake of environment. The environmental priorities have been changed now and it was the effect of historical summits such as the Montreal accord and Earth Summit in Rio, 1992.

DEFINITION OF SUSTAINABLE DEVELOPMENT

Sustainability is referred to be the long term health of global ecology, while "sustainable development" is a process hat enhance human and socio-economic well being in a long way. Presently, we are threatened by our own poisoning.

The World Commission sounded the most acceptable definition of sustainable development on Environment and Development (UN Sponsored Commission), as "development that meets the needs of the present without compromising the ability of future generations to meet their own needs". Mahatma Gandhi (India) was also very clear about the environment and sustainability, as he warned the developers " environment can save your need; can't your greed".

In fact, the present pattern of so called growth can not be sustained. We can have only hope on new pattern. The notion of development is taken to be much broader then a mere economic growth or may say gross domestic product, adopting:

PLATO'S THEORY OF UNSUSTAINABILITY

All other lands were surpassed by ours in goodness of soil, so that it was actually able at the period to support a large host which was exempt from the labours of husbandry... And, just as happens in small islands, what now remains compared with what then existed is like the skeleton of a sick man, all the fat and soft earth having wasted away, and only the bare framework of the land being left. But at that epoch the country was unimpaired, and for its mountains it had high arable hills, and in place of the swamps as they are now called, it contained plans full of rich soil; and it had much forestland in its mountains, of which there are visible signs even to this day. Moreover, it was enriched by the yearly rains from Zeus, which were not lost to it, as now, by flowing from the bare land into the sea; but the soil it had was deep, and therein it received the water, storing it up in the retentive loamy soil; and by drawing off into the hollows from the heights the water that was there absorbed, it provided all the various district with abundant supplies of springwater and streams, where shrines still remain even now, at the spots where the fountains formerly existed.

- the quality of life
- educational and cultural facilities,
- health and nutritional status
- access to resources and equity share of the resources
- freedom of having associations, deliver speeches and vanguarding the democratic norms

Measures of wealth and well being premised on the benefits of material consumption will need replacing with measures of resource sustainability and quality of life.

Sustainable development relies on ensuring that the local and global environmental resources, on which communities depend are maintained so that the resources can be passed to the next generation in intact. And, the major resources, which can be categorised, are

- Earth: soils, minerals, land and landforms
- Air: atmosphere, climate
- Energy: sun, energy supply, alternate energy
- Water: rain, water courses, water supply, rivers, sea
- Life: wildlife, trees, biological diversity, species

The key aim of sustainable development is to ensure intergenerational equity at the level best. It sets atleast four goals:

(1) economic efficiency in the use of scarce resources
(2) intragenerational equity – particularly equity of treatment between the developed and developing world, fairness in the terms of trade and other resource allocation systems,
(3) fairness to nature, treating other species as far as possible as partners I the shared biosphere, rather than as resources to be exploited; and
(4) resilience and robustness – survivability

CHAPTER 1

ENVIRONMENT AUDITING

1. ORIGIN AND NATURE

In fact, The UK and United States took a first step to understand Environmental Auditing. This process of environment audits originated from commercial response to natural requirement and not from the local authorities. Later, it resulted as a legislation, which made companies responsible for environmental loss they were causing. The US has adopted a principle "the polluter pays" so as to compensate the environmental loss. In order to avoid this liability and the companies took initiatives with regard to the legislation by way of conducting "performance review" and "compliance audit". During 70s and 80s of the last century, a number of anti-pollution laws and legislation came into light, in which Resource Conservation and Recovery Act (RCRA), the Clean Air Act and the Comprehensive Environmental and Liability Act (CERCLA) can be underlined. These regulations have forced the certain US companies to adopt obligatory environment auditing. Other companies have also adopted the environment auditing voluntarily as a sign of their environmental uprightness.

In the UK, a few large companies -mainly the British Petroleum-introduced guidelines of environmental auditing for the first time. When the US companies or subsidiaries started illuminating the state of affairs- environmental auditing" process and practices of their parent companies, it was widely disseminated. The concept of commercial environmental auditing has become broader in the years since its inception and now it is widely acceptable as a major tool for promoting a safe environment management, particularly for large scale companies. This commercial environmental auditing involves input and output analysis and call for an answer about the environmental impacts of the raw materials and final products brought into the factory and also about the impacts of the products and wastes that becomes apparent from the factory as a result of production and administrative processes thereof.

It was then, made applicable to the local authority sector in this country when "Environmental Charter for Local Government" of the Friends of the Earth came in light in the year 1989. By 1992, more or less half of the English and Welsh region local authorities stood for environment audit - completely or partly.

Recently the Government itself has expected local authorities to adopt environmental issues in a greater way and flourish public concern for green effects has precipitated local political responses.

Many authorities have produced environmental Statements and Charters, responding to government instructions in various spheres - waste or garbage disposal, pest control, building preservation. The local government has taken a much more synoptic view of the environmental than the traditional segmented delivery of environmental services. Environmental Auditing has become a vital means of converting aspiration to effective action. As experience of environment auditing is gained, shared and disseminated, it is becoming clear that they have significant role to play by encouraging the systematic incorporation of environmental perspectives into many aspects of policy and also by helping to precipitate new awareness and new priorities within the authority and without the authority both.

The pressure for changing the environmental scenario is also coming from "bottom up (that is from people as a voter and consumers) and top down (that is from international community) both. Perhaps, the most Earth Summit-Rio (1992) can be underlined as the top down influences which inspired the countries on the globe to review their environmental stand and to act effectively to save the earth with sustainable approach. It was aftermath of the Rio earth summit, the governments have produced their national strategy for sustainable development, which includes the policy, and programmes aimed at to promote geo-biodiersity and to establish green house emissions.

May not all the governments or the local authorities are getting in the Rio spirit with equal enthusiasm, nonetheless, there is an evidence of significant progress, which is showing their national commitment to change and upgrade the environmental situation to the possible extent.

European Union is the important catalysts for change (though the Asian countries are also playing the same role within their limits) who took initiatives in the field of pollution, energy and development. The UK have constituted Eco-Management and Audit Scheme (approved by the EEC in 1993) for the purposes of providing required help to industrial companies with a view to improve their environmental performances, the scheme, however, adopted by the department of environment, local government management board etc. and extended the scope to not only the internal environmental practices of local authorities but to their public services and policies as well. *In the year 1992, the union government of India announced the policy statement on environment with these words " industrial concerns and local bodies should feel that they have a responsibility for abatement of pollution. The procedure of an environmental statement will be soon introduced in local bodies, statutory authorities and companies to evaluate their policies, operations, and activities on the environment. An annual statement will help in identifying and focussing attention on areas of concern, practices that need to be changed and plans to deal with adverse effects. This will be extended to an environment audit. This measure will provide better information to the public.*

The Eco-Management and Audit Scheme, in his advisory capacity, provides assistance for environment auditing significantly, such as It provides formal management frame work for auditing and external validation as well. The EMAS emphasises "Internal" environment auditing as an effective management system and ignored the issue of external environment and its better quality.

Environment Auditing, in terms of auditing of the local areas/ authority does not have to cover each and every aspect of the environment or the local authorities operations. Nor does it have to be undertaken all at once. Auditing is a process - that can be conducted on whatever scale is appropriate to the needs and resources of the authority concerned, it does not require the extensive use of consultants. Some training and assistance may be necessary, but in most cases it may actually be more appropriate if the authorities own staff carries out auditing. Environment Auditing is therefore just as allowing of and useful for the small districts/or, vulnerable areas.

PRINCIPLE ELEMENTS OF AN ENVIRONMENT AUDIT

Environment audit has its two principal elements -

External Audit and Internal Audit.

External Audit commonly referred to as a report of the environment department. It involves study on the environmental conditions prevalent in the industry/local area. The external environment audit is therefore more about systematically arranged existing sources of information to compare and identification of the Gaps, which may sometimes can be filled or stepped by collaborating with the company, with the local community groups or other agencies. It sets the scene by assessing the actual quality of the produces and its effect on the local human environment.

While the Internal Audit assesses the policy and practices of the local authority/company in the following three facets -

(i) The review of Internal Practices
(ii) The Policy Impact Assessment
(iii) The Management Audit

(i) The review of internal practices - assesses the direct environmental impact of the activities of the organisation, the energy efficiency of its machines, vehicles, buildings, the impact of its purchasing, the recycling or the disposal of wastes. The review of practices is an important aspect of the audit, which is equally applicable to any organisation or industry.

The Policy Impact Assessment - The policy impact assessment is concerned with the environmental impact of the company or local authority in its specifically required role as motivator, regulator, enforcer, enabler, educator and service provider in terms of influencing the environment such as pollution control and landscape policies. Other social and economic impact may also be assessed if that is having environmental side-effects- e.g. housing policy and economic development etc. The policy impact assessment may therefore defined " asking their policy related questions which not

Table 1.1 : Potential costs and benefits of environmental auditing

Potential costs	*Potential benefits : (and needed)*	*Potential benefits :*
1. Commitment of resources to undertake the audit, including possible disruption of normal work 2. Extra costs of actions recommended by the audit 3. Increased liability, or public disappointment, if action is not taken 4. Threat to established patterns of work, priorities, personal empires (can be a benefit) 5. The unveiling of embarrassing actions on inactions (clearly a benefit).	1. Pinpoints actions to reduce contribution to global pollution e.g. the greenhouse effect 2. Reduced consumption of non-renewable raw materials e.g. fossil fuels 3. Improved quality of local air, water, soils, and wildlife habitats 4. Improved quality of the visual, aural and olfactory environment for local residents/workers/ visitors 5. Promotes, by example and publicity, environmental awareness amongst business and residents	6. Enables compliance with national and European laws and standards, and a consistent/structured response to new requirements 7. Reduction in some expenditure, for example for heating and lighting 8. Reduction in surprises, identifying matters needing attention, avoiding legal suits and/or public approbation 9. Improved public image and environmental accountability 10. Enhanced profile with business and government bodies 11. Independent verification of policies and procedures 12. Improves motivation amongst employees by giving an outlet to 'green' impulses and increasing clarity of purpose 13. Increased information transfer and better channels of communicaton between sections. 14. Improved environmental awareness amongst staff and councillors.

have been asked before and examining the policy related environmental interactions that were put in quite separate categories previously.

(ii) The Management Audit - The management audit stands where auditors assess whether the organisational structures, job descriptions, patterns of their responsibility and communication helped or hindered the environmental effectiveness.

Now it requires that the environmental audit should come to be observed as a regular corporate review, where every department of the authority or company makes their progress note towards the agreed targets. These targets may be incorporated in the action plans at different levels. The local authority/company could also produce an abridge annual statement by summarising their environmental performances so as to keep the same which may require by the government for taking corrective measures, if necessary. It must be informed to public and the political groups for debate on the future priorities of environment. The environment audit becomes, therefore, an institutionalised process of maintaining and gaining environment awareness of high order among the implementors, decision makers, policy planners, environmentalists, and public that can be triggered into an environmental thrust.

NEED OF THE ENVIRONMENT AUDIT

There are very practical reasons for environment auditing - geared, as outstanding ambition is becoming the order of the day. Local authorities/company's are increasingly being responsible for their impact on the environment. Government finds them as key players to launch sustainable development programmes. The moves towards strategic environmental assessment are now widely accepted throughout the globe forcefully. Local authority/company will soon find that without proper accounting of the environmental strategy their credibility with the funding or sharing bodies will be thinner. This will be treated as "Lack of an important ingredients".

Reliance on political green speech and professional green idea is no longer acceptable. The public, the environmental groups and the government as well, in fact, expects a greater objectivity and honesty in order to maintain environment consciousness and sustainable development at large. May be the local authorities or companies are not well equipped to respond to such challenges. The process of an environmental audit could hopefully assist them in making a start to achieve the target.

Aims of Environment Auditing

The environment audit is not confined by the remit of a particular department. It examines the problems of environment in every direction within a radius. This holistic view, of course, meant for breaking down barriers between different departments and different professions by adopting a corporate approach. The environment actually means different things to an engineer, to a manager, to a town planner, to an ecologist.

The holistic view includes dimensions of space and time. For example, on the local to global dimension it may very difficult to admit where exactly to stop auditing. Macro issues like global warming, ozone depletion, acid rain are relevant in, so far as, local actions have impact on them. Macro issues like specific product purchase; building management or species survival can not all be explored comprehensively within a limited budget and time. A sensitive question of an individual autonomy/responsibility may also arise if an individual feels their work unduly taken under microscope then he may resent and the suspicion or mistrust can definitely undermine the value of the audit. Therefore, it is always important to deploy or involve staff having vision of auditing priority.

Same arguments may apply to the time dimension. The holistic view emphasises the concept of revival approach where the whole life cycle of a product, from extraction to manufacture, to use to reuse or disuse, is placed under the environmental microscope. This approach is required by the Environment (Protection) Act for certain industrial processes, while local authority audits will normally have to rely on standard national formulations - for example, in terms of choices of building materials there are various studies on energy qualities and pollution implications exists which can be used.

It is also pertinent that the environmental audit should be open, objective and honest too. The audit will have a little value or no value if the defensive attitude of bureaucrats or departmental power games weaken the truthfulness of the results of the local authority audit. In an audit of a local area, the officers may be active in suppressing the clear evidences by misquoting that the other responsible departments did not communicate properly on environmental issues. It is therefore important to ensure that a high level support system is established for the audit at the outset from all relevant departments and committees, together with a firm commitment to act on the result in a appropriate way. It may not happen in the industry/company exactly what is said in terms of local areas auditing but the corporate executives may suppress the clear evidences and truthfulness of the environmental implications.

The another key goal of auditing is openness. Audit information necessarily be open and should be kept free available for the community people and environmental groups so as to raise debate on environmental issues largely in the community. In fact, the active involvement of the public and the environmental groups helps in validating the audit exercise and raise the profile of the local authority or companies.

Facts of the audit have a little value of themselves unless they are seen in the right perspective or context and their significance are appreciated. Knowing the level of SO_2-ClO_2-Pb-BHC or any other toxic emissions if there is nothing to replace or compare it with. So part purpose of the auditing is to evaluate variables against some kind of criteria or standard. A question may arise that whose criteria should be used? Should it be those of the environmental authority within the government, or a local politician or savers of the Earth?

It was made clear in the local agenda 21 and the Rio summit that " the community's interest should be served. The public should be invited/involved in selecting the criteria, which reflect the community's values. In some cases, environmental authority within the government sets the national standard. In the European countries - WHO, the European Union, the European countries have specified " acceptable" levels of pollutants - though they do not agree always. In Asian countries, there are various acceptable levels of pollutants. In India, Ministry of Environment & Forests and it's constituted body Central Pollution Control Board has set the norms and national acceptable standards/level of pollutants to be followed by the industries or by the local authorities. Though it can be more difficult to follow the norms exactly defined or set, while evaluating the environmental impact of an agency's regulatory. In the field of land use policy, there is no easy indicator of success or failure. Though the specific policy is likely to have environmental costs and benefits both. For example, a greenbelt may be taken as environmental asset in its own right - if its preservation of open countryside near a town. But, at the same time housing development may have overtake alternately in longer, car journey to work is probably bad or liability. Though the green belt attributes always "good", it is dangerous to apply predictable reactions rather than actually studying the broader environmental consequences of the policy.

Environment Auditing, from the beginning, aimed at as a means of change in positive direction and not just as monitoring exercises. When people change the policy and practices automatically changes. The audit organisation needs, therefore, to motivate people/authority and to look new, something different in their approaches or activities. It should try to impress fresh awareness and improved values. A good auditing process involves the relevant and effective training and it is more important than extensive monitoring exercises or a generous final audit report.

CONCLUSION

Local Authorities/Companies influence the environment through their own internal actions: waste reduction or recycle, transport, maintenance, energy management etc.

Local Authorities/Companies have long been involved in environment control and management in their role as planners, monitors, regulators and enforcers.

Local Authorities/Companies can influence public opinion by publicising and making available the information on a wide range of environmental issues.

Local Authorities/Companies should act as leaders or trendsetters in the best interest of the locality or local community people.

Local Authorities/Companies can influence environmental practices in the capacity of service providers or manufacturers.

Local Authorities/Companies can undertake environmental actions in a positive direction and can play a significant role as a major property owner and landowner.

Local Authorities/Companies can promote and support environmental improvements by way of financial actions such as by making provision of grants or investment of its funds.

Local Authorities/Companies can work locally and intervene where market forces are found indulged in promoting damaging activities. By way of this action at local level, sustainable development improvements in environmental quality can be ensured. Their involvement with the environmental practices is needed at a number of levels that is essential during auditing of the environment whether of companies or of local areas.

The auditing process is designed to ensure that the various dimensions of work are environmentally positive. In the long run the process will become just a normal part of the working practices of the local authority or company. Initially, however, resource constraints may limit the range of auditing. The philosophy behind the audit is to tackle first those elements, which can achieve a clear and high profile result than prepare the grounds. A progressive and continual approach may have benefits of learning and financial as well.

Recommendations

- All the local authorities that include companies also should undertake environmental auditing as a matter of course.
- The process of environmental auditing should be aimed at to be inclusive as possible it can be. That may include appropriate external agencies and the community by ensuring that the ownership of the environment audit is shared.
- The audit needs to be managed corporately with representation from all departments.
- The auditing should not be seen as the exercise not to be repeated but as a recurrent review process tied with the main programmes and policy plans that enables a regular view (to be followed easily) of overall environmental impact.
- The audit needs to link in to strategic environmental appraisal procedures within the company or authority and contribute directly towards sustainable development.
- Authorities should carefully consider the merits of installing environmental equipment's and management procedures followed by the set norms or standards of the regulatory body within the government.
- If audit starts without help, authorities/officers concerned should start with those element first that are easiest to follow or demonstrate the progress on and get strength derived from an initial effort for the furtherance of the audit in totality.
- The authorities of a company or an area should be aimed at to take initiatives to a broader community audit and not just a unit audit, as soon as possible.

CHAPTER 2

FRAMEWORK OF THE ENVIRONMENTAL AUDIT

INTRODUCTION

As it is earlier said that the environment auditing is a process and not an event. A process that is about people and attitudes about objectivity and technical procedure. We now discuss and examine

- the scope of audit
- the auditors
- audit tools
- community development and interests.

Each of these factors need to be considered "stated in detail", and consistent decisions taken, for the auditing process to fulfil aspirations.

While discussing the scope and purpose of the environmental study, two logical point of views may be discussed. Either the study or survey should make an attempt on synoptic review or environmental review of environmental quality or it should concentrate on just those aspects, which are answerable to policy or law. The earlier attempt has the baseline that the local authority is responsible for safeguarding the local environment and is having role to play in terms of monitoring the environment and awaring the community and the concerned organisation. A synoptic review stands on the fundamentals, identifying the key indicators of environmental quality for each of the elements. For example, in terms of the land, it examines the quality of land and soil - its stability, fertility, erosion problems, pollution or loss of strength. In terms of wildlife and biodiversity it examines the level of locally rare species or endangered species, the extent and quality of different habitats and put pressures that may lead to change. In terms of perception, it might study the general quality of visual appearance of town and land and residents perceptions as well.

A policy oriented environmental study takes a much more limited and pragmatic view. Basically its aim is to collect information, which allows the auditors to assess the success or failure of policy to achieve the specified goal. If there is policy concerned with reducing waste/illegal dump, then is it reduced actually? If there is policy to safeguard trees and woodlands, then has the rate of tree loss stopped or slowed down or still have losses more than compensated for (by planting more trees). If there is a provision to check abnormal obstruction, then is it decreased? By this approach, the auditors, in fact, monitor the effectiveness of policy.

Approaches to Policy Assessment

Four basic ways are identified and in practice, for evaluating the policies and practices of the authority. The ways are having difficult implications for the auditing process and also for the resource level needed. That is

- measuring the actual impact of policies on the ground level
- estimating the impact of policy on every elements of environment
- comparing the policy with best practices, if it exists elsewhere
- assessing the effectiveness of the management systems that reflects the policy clearly.

Measuring the real impact links with the policy oriented environmental study; identifying indicators provide a measure of policy success or failure. If the objective or target for each indicator is stated in detail, the task of assessment will be easier. It is desirable to evaluate performance clearly by examining actual outcomes. In the long run it may be essential that all spheres of the policy be monitored and reviewed the procedures carefully, but in the short run it is not practicable. The reality is that only a limited number of key variables can be systematically monitored.

The professional's services can be utilised for systematic estimate of policy impacts through the use of matrix. These matrix - checks and balances- are valuable and contributes to the monitoring process. It may, however, hit or miss and there may be a possibility that a potential and critical impact elude the notice because these impacts are not on the top of the political agenda.

In this state of affairs the other approaches become more important. The common way of assessing the content of practices and policies is to compare them with some influencing qualities. Checklists of the model policies can be prepared by environmental bodies to meet the purpose. Government provides guidelines on planning policy. There are evidences of academic studies of innovative authorities in the country where the overseas countries provide another source. All these can be used to assess productively where the authority/company be spreadout to view a good practice.

The significant advantage of a straightforward comparison with good practices is that it is quick, directional and pointing the way forward. The limitation is that it does not assess how far the policy is being translated into action. In many circumstances the content of policy may be assessed with less significance than the company/authority ability in successfully implementation of the policy.

Assessing the management system focuses on human behaviour and organisational structures rather than policy content or direct environmental impact. The auditor's questionnaire may include the question "Are there expressly stated environmental objectives"? Are there programmes or strategies for achieving them? Are the responsibilities clearly defined for implementing the programme? Are there effective monitoring and review procedures? By applying this approach it is assumed that people involvement matters most and whether they have appropriate goals and values and also whether the institutional framework permits them to be effective. Hence the key to improved performance is organisational culture and well awareness of the staff. If the attitude and the framework are taken in the right direction, the right decisions will certainly flow.

FRAMEWORK FOR A COMPREHENSIVE AUDIT

It needs no further explanation that elements of the audit can not be tackled separately. It must be considered in relation to each other elements.

The level or quality of primary stock represents a baseline measure of sustainability and audit needs to assess the current pattern of sustainability with its extent. This is important in terms of Air, Water and Earth quality particularly, the level of greenhouse emissions, and the maintenance of biodiversity. It also provides a measure of the quality of environment-in general actually availed by the people in the area.

Secondary indicators provide definite or available means of progress to be assessed. For example, the level of traffic is not itself a measure of absolute sustainability or quality but may act as a useful substitute for transport energy consumption and CO_2 emissions.

Third indicators are specific to the particular agency that provides a direct measure of policy achievement with the reservation about cumulative or indirect impacts. These specific assessments are central to the policy impact assessment - as a conclusive test of policy implementation. Selected indicators should be identified when the policies are devised and approved.

Generally the indicators should be selected which either have global significance, as defined by the internationally accepted organisations or have local importance as defined by local pressure groups, environmental groups, political groups and the public.

The indicators require standards or criteria to make measuring tools or acts meaningful. These are normally defined in the policy documents.

The first part of Auditing

While initiating the new system, comprehensive approach from the beginning is ideal, though it can be difficult and expensive. It may have demanded much change quickly and it requires heavy investment of time and money both. So it is always advisable to have a phase approach, which can be helpful in sparing more time to experiment and persuade with the matters.

Each local authority/company will have its own priority, but the easiest and economical element, to tackle first, is to review of internal practices. On each level it creates a sense to deliver such things and it is always good to be seen, to be putting your own house in order before making lecture to others on what they should do on the particular issue. On the organisational level it is useful because review of internal practices will effect all the departments, raising environmental awareness across the boundary. In addition, it gives definite results from the review of internal practices. It is less easy to specify whether environmental study or policy impact assessment should come first. In fact, the information provided by the environmental authority is subsequently used in the policy impact assessment; however it could be premature to identify what information is required without first considering the policy impact assessment. It may be better choice to deal with both in following the other topic.

In absence of overview of the issues and their interrelationships, problems may come up to deal with. While on course of action, policy impact to be tackled separately, department by department. Apart from the obvious dangers of self-validation, such a procedure falls into the device of assuming the environment and it became departmentalised. But the statutory requirements that inclined to define departments can get in the way of environment conscious policies or even the recognition of an environment problem. Conclusively, the policy impact assessment/environmental study should be interdepartmental, and where possible interagency may be allowed collaborative solutions to emerge. It means that there are considerable benefits from a comprehensive approach, across all issues.

We find, therefore, there is no perfect answer. It will depend on local circumstances. The vital thing is to put on the process, in a very carefully and considered way, as soon as possible.

Identifying the auditors

The issue of audit - whether having comprehensive or incremental approach- is influenced strongly by the question, who should be the auditor? There may be three possible basic answers:

- The audit could be done by outside consultants, with assistance of the internal staff.
- It could be done internally with external experts consulted on specific or/and limited issues.
- Or, it could be done by internal staff and consultants working as one behind another.

The first option is main driving force while the others provide guidance, training and validation of the audit process. It will depend on the capability and ambitions of the authority or the company, while identifying the auditors.

Consultants

In this age of privatisation and contracting out the work, the services of consultants have the advantage of relative objectivity. It reflects the operations in an impartial way and provides new perspectives on issues and opportunities. It has also has the benefit of choosing the right and expert consultants. For small companies or authorities where staff is limited in terms of size and knowledge, *it would always better and practical way to use the services of consultant for the purposes of audit. In terms of a company audit, the use of consultants may be the only practical way of doing an audit.*

The disadvantages of environment audit done by the external consultant auditor can be underlined ideologically such as the audit may be treated something which is imposed upon staff rather than being a process they own, and as a result their motivation to implement the audits recommendation is undermined. They may be disappointed if the audit report suggests the staff to take corrective measures immediately on the basis of the information-documentary or verbal - provided to them by the same staff. The results may therefore be rather predictable and not issue solving. These are the risks of using consultants simply to defuse the political or administrative pressures, and make an empty environmental gesture, rather than enabling the follow through the issues in an ongoing auditing process.

In-house Auditing

If the in-house staff carries out environment auditing, the topside of environmental issue is likely to be true. While on the process of auditing it is supposed that the in-house staff will themselves have to confront the issues and make proposals for further improvements. Active involvement may enhance their sense of belongings in the entire process of environment auditing and their motivation too. The means of implementation and subsequent monitoring will be under their control.

But the risk is here that the existing staff will not take steps for thorough reform. They may find the audit as a means of validating current practices. Lack of wider experience and lack of time which is required for the audit, could make the exercise restrained and even tokenistic.

The review should be corporate by ensuring contribution of every department. In this context the status of the auditor becomes crucial. The audit conclusions may be treated with doubt if they derive from one department or are relatively junior they will not have sufficient authority.

Combining the external auditor and in-house staff

Relying on internal staff may be the question of reliability and it weakens the audit obviously. The solution is, therefore, to have a combine operation for audit, by deploying the consultants for specific tasks e.g. guidance on scope and management of the audit, training on assessment techniques and best practices, and verification of the processes. In-house staff would remain in control and carry out most of the audit tasks. If the right consultants were chosen, this approach could draw on the strong points and will be more beneficial. The consultants are supposed to have in depth understanding of the auditing processes and knowledge of best environmental practices and also the ability of communicating their knowledge. So, it values if consultants are deployed but it should not be the main driving force of the entire process of auditing.

Community Involvement

The Public

The public may not be helpful in the process of auditing but they have certain roles in terms of :

- Sources of information on attitudes values and priorities
- Source of information on the state of the environment, specifically local problems and conditions of environment
- Means of implementing change by improved environmental awareness, altered behaviour, pressure put on organisation which pollutes the environment.

Public attitudes, values and priorities are normally reflected to the employed professionals. Responses gained through the leaflets and meets are similarly self-selected and liable to prejudice. So if a fair representation of community attitudes is desired, a sample survey is essentially be conducted carefully.

Environmental concern may well be highlighted by public involvement which usually given low priority by the auditors. However, it may not be wise to simply follow the public view in everything in the light of global environmental guidelines.

The need to communicate with the public puts obligations on the auditors to make auditing accessible. The key documents may not likely to be learned technical papers but short, systematically illustrated reports that set out the issues clearly. If material is put on a computer database and given public access, more detailed level of information may be available to those who are interested to know.

The local politicians may be drawn into the process of auditing in the following places:

- In a principle decision that the audit exercise is to be undertaken and also in decision of the resource level in general.
- In agreeing the scope that the audit should be broader narrow to what extent, the time scale and relationship with other policy review processes in details.
- In publicising as the key issues are explored, explained and encouraged required participation
- In reviewing the emerging results of the audit
- In deciding the implementation of the recommendations which may include allocation of funds
- In subsequently maintaining support and resources necessary to monitor progress and sustaining ongoing auditing.

Other possible involvement

The state of environment is of value not only to the local authority or company but also to other organisations with statutory roles in environment monitoring control. The value of environmental study may be stated undermined if information from other sources is not taken or available. These includes Pollution control bodies, authorities on Air-Water or Natural resources, and beyond the official bodies - NGO's having expert members in the respective fields of environment, having vast knowledge of the area may be able to contribute much data to be used. They may also have points of criticism of the authority or company, giving a different point of view on a problem, which may be useful in resolving the environmental problems.

Tools for Auditing

While viewing the technical approaches, the following tools might be employed:

- The geographical database
- The policy impact matrix
- Better audit practice checklist
- Consistent analysis

The geographical database

The Department of Environment in India within the MoE&F, Union Government provides a baseline on the basis of which progress can be judged, and it is desirable to be able to update the database to a format. A computerised database, capable of manipulating space patterns and accessible to the public as well as to staff, may be helpful. It may be achieved by using computer-aided design (CAD) package in combination with standard database and official survey based data, or more ambitiously using geographical information systems. This can be valuable for other local authority/company tasks as well in terms of development control measures. Both these system have considerable potential subject to put up the cost and training requirements of the audit.

Whether a computerised system is used or not, the department of environment report itself is likely to be used widely and widely quoted. It is a saleable commodity for developers and other enterprises. It is essential. It communicates accurately and effectively. Since, the data will be regularly updated and extended a loose-leaf system could be considered which allows to include the annual supplements.

The Policy Impact Matrix

To bridge the gap between the departments and to ensure consistency, matrix system is to be considered as one way. It is designed to systematise the process of estimating impacts, which provides a framework for recording professional decisions. The format is straightforward-moves on the policy areas as one fixed reference lines for the measurement while environmental quality aspects are listed across the other reference lines. It is possible to breakdown both these set references into much more specific and numerous elements, in order to differentiate environmental effects more precisely. The matrix encourages a systematic and comprehensive approach, at least every possible impact is briefly considered. The technique allows the auditors to establish procedures from a fair set of comparisons.

There may be some problems when applying matrix approach from several angles. In the first it runs the risk of being unduly burdensome, because in order to reflect the real aspects of the policy or environmental interactions, the size of the matrix to be large, the process of recording the nature of an effect, criteria for assessing and estimating the significance becomes a difficult and confusing task-the auditors may become drawn into the tasks of completing the matrix rather than the task of achieving better policies. The second problem originates from the first, as it is difficult to achieve consistency in the matrix unless the same small group of people tackle the whole exercise and that excludes others from the ownership of the process. The third problem is the most fundamental as the matrix does not allow to the effect of combining policies. There are two important ways in which this has significance; by cumulative impact - where the overall effect is greater than the sum of the parts and by interactive impact - where a particular policy has different effects depending on which it is combined with other policies.

So the matrix is not a remedy for the auditor in trouble. If relied, it is likely to lead to significant inaccuracies and may sometimes be dangerously misleading.

The matrix idea definitely has merit as an act of prompting, not a restrictive measures. It can provide a standard impact checklist for use by the different groups. It should be kept broad and generalised, so that it can be tackled quickly, and used as a scoping exercise to help set priorities. It may point out the matters which was overlooked in previous attempt.

CHECKLIST FOR GOOD PRACTICES

While matrix approach is about estimating the actual impact of policies on each element of environmental stock, the alternative approach is to look at the nature of policies. This can be done by comparing them with model policies traced from external sources such as :

- Statutory environmental bodies lying with several government departments,
- Non-governmental environmental organisations,
- Government department who creates and regulates policy/ planning/guidance on environment,
- Other local authorities who are known to be at the forefront on the issues of wildlife, energy, strategic sustainable development planning, environmental planning, forestry, landuse planning etc.,
- Other countries
- Relevant literature, practical recommendations of the environmental bodies.

In fact, it is a part of professional's duty to keep their updated checklist with emerging good practice from such sources in their field of action or interest. List of model policies can be used then as templates against which to judge the current performance of the local authority or company. (A highlight is given to provide the possible comprehensive checklist which could act as to begin with the auditing.

The advantages of this approach are that it is simple, direct and points the ways forward. It does rely on honesty having no

Table 2.1 : Framework of aduit elements and tasks

Main elements		*Audit task*	*Examples*
SoE	1.	Primary indicators Monitors the basic quantity of environmental stock	• Air quality • Water quality • Mineral reserves (local) • Townscape quality
SoE/PIA	2.	Secondary indicatos Measure basic quality by proxy and assess the general effectiveness of policy	• Preservation of SSSIs • Trafficlevels (proxy for transport energy use and CO_2 emissions) • Rate of recyling in the community
SoE/PIA	3.	Tertiary indicators Assess the *direct* effects or effectiveness of policy	• Have out-of-town retail proposals been blocked successfully ? • Have planned cycle routes been constructed?
PIA	4.	Policy impact guesstimating Evaluating current/proposed policies' likely impact on every element of environmental stock	• Use of a matrix approach–reliant on informed judgement
PIA	5.	Comparison with good practice Assess scope and content of policy against external yardstick	• cf English Nature recommended policies • cf DoE PPGs
PIA/RIP	6.	Management audit Assess the effectiveness of the environmental management systems and service delivery.	• Review objectives, programmes, means of implementation. monitoring systems etc.
RIP	7.	In-house indicators Monitor actual progress towards in house objectives	• Proportion of paper recycled • Energy used in the buildings • Water consumption of LA

compassion. The auditor has to distinguish between stated objectives of policy and the policy actually being implemented. Another problem that checklists - in the matrix approach- usually fail to address the policy interaction. Where the environmental effects of a policy are dependent on what happens, in another field of activity, the presumptuous behaviour of excellence may be simple.

Network Analysis and Consistency Analysis

When dealing with strategic land use / industrial / energy / transport policies such as residential and commercial location policy, educational provision, density issues, traffic planning, transport-energy-industry policy, green belt, urban form, rural settlement policies and economic regeneration strategies, the problems of complex systems arises prominently. A consistence strategy ensures these varied interests are mutually supporting - not undermining-each other in relation to environmental goals. Consistency is also applying to all spectrum of the audit process. It is well recognisable that separate areas of pollution control, waste collection and disposal, sewage treatment, water supply and drainage, energy production, mineral exploitation and recycling on the surface-all have intense implications for each other and relate also to issues of urban form and development patterns.

Few audits have even tried to manage these complexed questions. To that extent most PIA's so far have been less than adequate. To solve these tasks, the overseas countries used Analysis of Interconnected decision Areas (AIDA) approach. AIDA, actually, provides a kit of simple tools-linkages diagrams, matrices and flow charts-capable of shaping the complexities and provides a framework for policy evaluation.

The process of analysing connections should be tackled systematically, by linking policy areas of common environmental impacts. In terms of land use and transport example (which can be applied in auditing of corporate sector or others by making some amendments or corrections), the suggested stages are :

- ***Identifying a cluster of linked decision areas*** - this may be useful in consideration of all types of connections between policies, in gaining the acceptance of different departments

Table 2.2 : Mechanisms for conducting an audit

	In-house team (existing staff)	*In-house team (new specialist sttaff)*	*External consultant*	*Combined in-house and consultants*
Cost	Cheapest	Most expensive but could be the most cost-effective long term	More expensive than in-house team	More expensive than in-house team
Overall environmental expertise	May not be comprehensive	Excellent and LA controls what it gets	Excellent (if right consultant selected)	Excellent (if right consultant selected)
Knowledge of authority	Excellent	Limited at first but will grow	Likely to be nil or very limited	Good
Knowledge of area	Excellent	Limited at first but will grow	Depends on firm's experience but could be nill or limited	Good
Objectivity	Bound to be affected by LA culture	Good to excellent	Excellent	Good
Ease of management and control	Good	Good	Not as easy as using in-house team	Not as easy as using in-house team
Commitment to long-term implementation	Depends on other duties and priorities	Excellent	Non-existent	Mixed
Continuous involvement in monitoring and review	Depends on other duties and priorities	Excellent	Non-existtent	Mixed

A SIMPLIFIED APPROACH TO THE ASSESSMENT

For each set of Issues (policies, practices or environmental topics) tackle the assessment in the following sequence:

1. General introduction and review.
2. Current status of the issue with regard to the local authority.
3. Statutory obligations: best practice elsewhere; possible standards, targets and indicators.
4. The assessment of 2 in relation to 3.
5. options for action.
6. Conclusions and recommendations.

and agencies that our policies are linked and it requires collaborative evaluation.

- ***Identifying relevant linking issues or goals*** - it helps in focussing the debate on the nature of policy interaction.
- ***Preparing a compatibility matrix*** - it helps in concentrating on the real policies i.e. those that are being implemented, rather than any dull general statements of policy which may really be no more than aspirations. It gives such clarity on important issues. The matrix is liable to include factors at different levels - ranging from goals and objectives to specific proposals.
- ***Analysing compatibility*** - it helps in identifying different degrees of compatibility and the key reasons for reaching to the final answer. It reduces the negative effects on the key environmental goals.
- ***Drawing conclusions and making recommendations*** - It helps in stating conclusions and initial recommendations in relation to each of the policies and form a basis of a revised package of policies.

Recommendations

This chapter has tried to discuss the key choices, which need to be made in order to audit progress. It may be observed in this

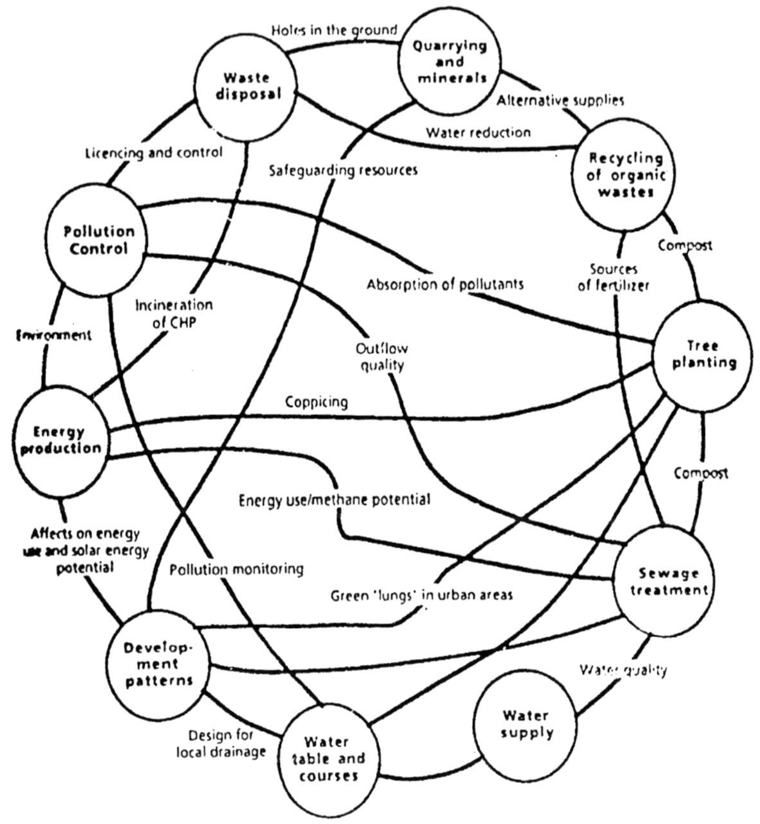

Fig. 2.1 : Network analysis of diverse decision areas

chapter that there is no single and right way to proceed towards audit. There are several ways to proceed and it depends on the context and purpose of the audit. Hence, the recommendations are:

- The choice of the environmental authority reflects the priorities and values of the community and of the authority or company. Public participation or involvement is desirable.
- It is always better to tackle the "Review of Internal Practices"- an evaluation of the environmental impact of an agencies

own operations and practices, so that the authority may be viewed to be "creator to put its own house in order".

- It must be ensured that the scope and orientation of the environment study is such as to give answers on policy impact.
- Scope of the policy and service activities should be carefully examined with a comprehensive checklist of good practices.
- The audit should be undertaken as a in-house activity to the possible extent, so that the officers feel their integrated and total involvement in the process and learning opportunity through it. However, external consultants independently or jointly with the "in-house officers" are needed to provide training and expert ideas where the "in-house officers" is missing in terms of objective validation of the audit process.
- The style and presentation of audit material should always be accessible / transparent and to encourage the effective involvement of social activists, the public and different department of the house in the entire process of auditing.
- Detailed infrastructure and environmental data should be available to developers to facilitate their involvement in planning for sustainability. It may be provided on cost basis.
- Where policies reciprocally active with each other or have cumulative impacts analysis (consistent) should be made use of. Collaborative exercises are essential to recognised and act on shared issues - where several different sections or agencies are involved with their own agenda.

CHAPTER 3

ELEMENTS OF ENVIRONMENTAL STOCK

CATEGORIES OF ENVIRONMENTAL STOCK

In order to achieve sustainability, the basic condition that has to be satisfied is constant environmental stock in economic terms. Capital stock can be measured in terms of amount and quality of sources such as the quality of soils, the pureness of the air, the level of the energy resources, the quality of energy resources, the cleanliness or quality of water and/or the existence of diverse habitats. The in-house authority monitors stock quantity through the environmental study reports. They can assess the impact of policy on the environmental stock. If the authorities have desire towards a synoptic view of the environment, the elemental categories provide a starting point for a checklist, which can be extended to incorporate manmade stock and natural stock as well. The elemental categories are :

- Global ecology
- Natural resources (economically)
- Human environmental quality

Global Ecology : is concerned with atmospheric and climatic stability primarily and also with a conservation of biodiversity. Key variables responsive to be influenced by the authority, is CO_2 emissions and wild life habitats.

Natural Resources : is concerned with appropriate (economic) use and conservation of our resources of air, water, the land, and its minerals.

Human Environmental Quality : contains the manmade stock of buildings, infrastructure and adapted landscape in terms of appropriate space for human activity and its perceived better quality.

Elements of environmental stock

Component (element)	Key objectives for in-house authority
Global ecology	
1. Atmosphere and Climate	reduce $_{CO2}$ emissions reduce energy use in buildings / transport substitute renewables for natural fuel extracted from ground increase CO_2 absorbent
B. Natural Resources	
1. Air	maintain/enhance local air quality
2. Water	improve quality of water channels/ bodies protect water supplies
3. Land	maintain/enhance soil fertility protect from erosion and contamination
4. Minerals/energy resources	reduce consumption of non-renewables protect potential for renewables
Human Environment	
1. Buildings	availability and renewal of appropriate residential, social or commercial built space in convenient/readily available locations
2. Infrastructure	provision and renewal of necessary/ safe transport and service infrastructure
3. Open space	provision/renewal and cleanliness of accessible and appropriate open space
4. Cultural heritage	safeguard archaeological assets, historic monuments, good architecture, attractive townscapes and landscapes
5. Aesthetic quality	enhance perceived environmental quality in terms of sight, association, sound, smell and touch etc.

Degree of sustainability

Strong Sustainability - *it refers to factors risky to global issues. CO2 emissions and deforestation may be given as example. This is sustainability of high order.*

Weak Sustainability- *it refers more localised concerns. This is sustainability of comparatively low order.*

Architectural heritage or open space - this may be defined as sustainability of low order.

It may be useful, in particular, to analyse the global significance of local actions by applying these criteria :

Are emissions, mainly CO_2, NO_2, SO_2, NH_4, O_3 affected?
Are non-renewable resources conserved-consumed or substituted?
Is the recoverable/absorbing capacity of the biosphere degraded or enhanced?
Are rare or endangered or vulnerable habitats damaged or threatened?

Capital and constant stock

Those components, which possess such crucial value that they should be preserved at all costs, are capital or critical environmental stock, as critical to sustainability of our natural/man-made heritage. It is usually applied to specific sites or areas, such as sites of special scientific interest or outstanding conservation areas. Threats to any critical element of environmental stock would cause justifiable "strong public protest".

And, it is natural that the environment people or agencies are often to see their own interests reflected in the identified critical environmental stock irrespective of that their demand may not necessarily and always right. The alternative defence of constant and transferable environmental stock may be more appropriate in some cases. It is an altogether more perceptive concept. Constant stock might also be best for habitat preservation where natural plant succession and/or human impact can lead to quite rapid change on any particular site. The important thing is the general survival and enhancement of that habitat over a wider area.

Constant stock should therefore not be seen as a weaker level of protection than critical stock. It is appropriate and depends on different circumstances.

Transferability and Substitution

Distinguishing between different kinds of transfer or substitution:

Replacement of " like with likeables" in terms of maintaining overall stock levels. In case one area where tree grows are replaced by another area / neighbouring site due to felling may be defined as replacement of like with likeables.

Substitution of one resource to another: In case the use of non-renewable mineral reserve may be compensated by an equivalent increase in long term capacity to recycle the material is substitution of one resource to another.

Transfer or trade across boundaries. In most instances, local boundaries do not represent autonomous geographical or economic units. Here it will be necessary to see the locality as part of the wider region.

Environmental capacity, standard and Capacity

Unless objectives are set against which the performance of the environment and the effectiveness of the policy in producing better environment, can be assessed, there is a little value of monitoring the environmental impact. Such objectives come in the forms of;

1. Desired direction of change
2. Standards
3. Targets
4. Environmental capacity.

These objectives, in fact provide criteria for evaluation which includes target and levels set by the government for emissions and homegrown strategies, determined by the local processes, documented in the development plans, transport, wildlife strategy etc.

Table 3.1 : Elements of environmental stock

Elements	*Key objectives for Local Authorities/ Industries*
A. Global ecology	
1. Atmosphere and climate	Reduce CO_2 emissions : – reduce energy use in buildings – reduce energy use in transport – substitute renewables for fossil fuels Increase CO_2 absorption
2. Biodiversity	Conserve extent and variety of wildlife habitats Protect rare or vulnerable species
B. Natural resources	
3. Air	Maintain/enhance local air quality
4. Water	Improve quality of water courses and bodies Protect water supplies
5. Land	Maintain/enhance soil fertility Protect from erosion and contamination
6. Mineral and energy resources	Reduce consumption of non renewables Protect potential for renewables
C. Local human environment	
7. Buildings	Availability and renewable of appropriate residential, social and commercial built space in convenient/ accessible locations
8. Infrastructure	Provision and renewal of necessary/ safe transport and service infrastructure
9. Open space	Provision/renewal and cleanliness of accessible and appropriate open space
10. Aesthetic quality	Enhance perceived environmental quality in terms of sight, sound, smell, touch and association.
11. Cultural heritage	Safeguard archaeological remains, historic monuments, good architecture, attractive townscapes and landscapes.

Summary of chapter coverage relating to the audit

Audit Topics	*State of the Environment survey*	*Policy Impact Assessment*	*Review of internal practices*
Nature conservation	●	●	○
Energy	○	●	●
Transport	●	●	○
Planning	●	●	○
Aesthetics	●	●	○
Pollution	●	●	○
Waste and recylcing	○	●	○
Purchasing		●	●
Community awareness		●	○

Key
- [●] Prime areas covered
- [○] Secondary areas touched on
- [ ] Marginal significance

Capacity Planning

When dealing with the concept of environmental capacity, we should be well aware that environmental stocks do not emerge from natural laws. These are policy oriented and depend on the views and preferences of policy makers. These are putting on a test to search for absolute environmental capacity and use it to defend development pressures.

It is important to keep the concept of environmental capacity as an open and continual process. The first and foremost point to make is that the capacity planning is not entirely new. And, the second point is that there are various different types of environmental capacity, which includes the built environment and natural environment both. Even, the human environment can also be laid out at full strength. It can be therefore classified in the following divisions:

Built Environment - capacity of building stock, capacity of infrastructure, capacity of aesthetic and cultural heritage

Natural Environment - capacity of wildlife habitats/landscapes, capacity of natural resource base absorbent capacity and resource capacity

Human Environment - capacity in terms of social impact and traffic impact as well.

These capacities are changeable and open to manage the situation.

Natural Resource Capacities

While planning natural resource capacity, two components need to be discovered : absorbent capacity and supply capacity. The capacity of local air, water and land resources, and to absorb the emissions and effluents of human activity without compromising their origin or human health is in the centre of sustainability. One main task of the monitoring is to assess progress against quality standards. Statutory bodies (in India- Central pollution Control Board -CPCB), however, do not define the carrying capacity for an area, they emphasises on "pollution control and reduction" in the light of the various environment protection acts. The theory is that the development is fine but the developer must prove themselves "non-polluter".

The argument over the supply and the resources is different. Carrying capacity strongly suggests that development in any region should occur only at the rate that can be supported by local supplies of stone, gravel, clay and may be of energy as well.

But the reality is that region's trade in minerals with each other increasingly, across regional boundary even national boundaries in some cases. Some area will be notably known as importer and some one as exporters. So, it is difficult to see any socio-economic or political mileage while linking population and development directly to local resources for being aware of future.

Infrastructure capacity

Probably, the infrastructure capacity is the least mutable. The sewers, water mains, roads etc. have a certain capacity and are subject to centralised control.

Building stock capacity

The building stock capacity in meant for housing and correlated with population. Increasing population in urban areas is because of unemployment problem and undeveloped conditions in rural areas. It gives undue pressure on the environmental conditions and developmental processes. And, the capacity may gradually become change under market pressures and may be reflected in applications for extensions, divisions, and change of use and renewal.

Wildlife Habitats and Fragile Landscapes

The wildlife habitats and fragile landscapes are vulnerable to excessive human activity. The impact range is broad which includes either agriculture, quarrying or any developmental work or traffic, walkers, tourists' etc. In this state of affairs of multiple and different impacts, it is difficult to find out a common dividing line in terms of capacity for what. When one type of impact is prevailing then the capacity of a habitat or landscape to deal effectively with human invading can be changed by the management by changing the location of car park or creating a path or otherwise. In some circumstances the natural ecology itself regenerates, discovers new state of composures or invading new areas. When we deal with a living fluid situation, it is not easy to define an absolute capacity. It might be better to think in terms of beginning of impact, which demand action.

Non-renewable Resources

Most of the environmental stock have national or local criteria which can be chosen by the authority / company and that may be adapted as per demand or need.

In fact, there are no adequate guidelines of the government whether promote the concept of mineral conservation or permit to make out the meaning of that to the local level. Where resources such as oil, stone, gravels are freely traded between different areas of the country; the relevance of locality is too little. Sustainable exploitation of these minerals needs to be defined at national level but in the face of national need or increasing commercial demand.

If a sustainable policy were developed it could be based on around the principle of constant capital stock and critical measure (in terms of oil and natural gas) might be the extent of proven recoverable reserves. The rate of consumption would need to be reduced and the shortfall made up of increased energy efficiency or substitution by other renewables (in the form of fuel), if this was falling because consumption was exceeding the rate of discovery. Loss of non-renewable resources should be compensated by increased capacity to use those resources efficiently, so that the rate of loss will subsequently fall.

In other words the critical measure of sustainability is the longevity of reserves. The level of proven recoverable resources divided by the rate of extraction per annum should be maintained or improve year and year as extraction rates fall and new reserves are found.

This is a delicate measure. It recognises the process of mineral discovery rather than an arbitrary limit; it provides an incentive to energy efficiency measures and to substitution both. It does not prevent economic growth but shifts its emphasis.

It may be worthwhile to monitor such trends before auditing. Meanwhile the auditing should at least try to identify

- the level of reserves locally,
- the rate of exploitation of those reserves,
- the number of planning permissions being granted whether rising or falling,
- the extent to which the policy is moving to conserve scarce resources.

Recommendations

While measuring the national and international efforts towards sustainable development, the audit should be adopted as a holistic and comprehensive approach to the environment.

The environmental and economic success should be decided by assessing the levels and quality of environmental stock itself.

The elements of environmental stock - in its varying nature- should be reflected clearly whether global or local, critical or constant stock with a recognition that substitution and transfers of stock are appropriate in some spheres.

Standards, targets, capacities and thresholds should be used in appropriate context to assess progress and also to point out the progress needs to be made.

KEY POLICY AREAS WHILE AUDITING

Succeeding chapters are related to the realities of environment audit. The key environmental issues and the constraints and limitations likely to be met, are discussed. It deals with the topics related to the areas of expertise and specific responsibilities within local authorities, industries or companies. It provides guidance on auditing, which is designed around stages of the auditing process-environment management, policy impact analysis and review of practices.

The succeeding chapters provide a comprehensive guidance to the important issues. The environmental issues are:

- NATURE CONSERVATION,
- ENERGY,
- TRANSPORT,
- LAND USE PLANNING,
- AESTHETICS AND CONSERVATION,
- POLLUTION CONTROL,
- WASTE AND RECYCLING
- PURCHASING,
- COMMUNITY AWARENESS.

Chapter 4

NATURE CONSERVATION

Nature conservation is one of the major environmental issues, which is institutionalised in the work of local government. It is duty of the local authority to protect and promote the conservation of wildlife and habitats. While fulfilling the duty, a number of related issues seems to be addressed such as landscape, land use, natural resources, access, education, agriculture, woodlands, urban parks, local authority as well as community resources etc.

The conservation of a single local species or the protection of habitat has a role to play in maintaining global biological diversity. The issue of nature conservation can not be separated with the biological diversity, in the context of sustainable development.

Bio-diversity means the variety of life in the world, which can be termed as genes, species and eco-systems. A primary objective of sustainable development is to maintain viable populations of species, preferably in their native habitats and not in the farms-zoos-genetic banks. On the global scale, it is very difficult to assess diversity so the focus of attention is on nations, regions and local areas. This has been already been recognised in the convention of bio-diversity at Rio Earth Summit and the government of different countries signed the memorandum of understanding. The nation or local authority should kept the task manageable by way of focussing on nature conservation issues of local importance keeping an eye on the larger context.

This chapter deals with a framework for tackling this challenge and suggests the way of progress on the separate issues within the overall audit strategy.

Priorities

Nature conservation includes two inter related and mutually supported themes-wildlife and habitat. The wildlife and/or living components of an environment can not be separated from the habitats in which they live. Therefore, the issue of habitat conservation arises. In fact, protection and enhancement of habitats and wildlife is a part of planning process that has to be made by the local authority and should form the basis of the auditing initiatives. When evaluating the impacts of the local authority's policy and practices, the role of planning will be emphasised.

If the habitats have been protected from damages once, the complete range of other skills of the local authority will come in to role of active play. Protected areas will require then a likeable management and careful improvement. And, people will need to be informed and educated about the resources. It has to be coordination with other policy areas in terms of improving access, ensuring well-suited goals, and monitoring to assess change in environment. These are the important issues for environmental auditing. Hence, the basic fundamental reason is to be protected the local environment by the local actions. The nature conservation task before the local authority has to be implemented in the light of "nature conservation is about people and their environment".

SUPPORT LEVEL

The question of nature conservation now, has wide ranging support and pressure, as it was not seen before. It is only because of growing awareness of the people and "political will" (e.g. in depth understanding about the nature conservation at political level). The Ministry of Environment and Forests (Government of India) has enacted the Wildlife (protection) Act, 72 aimed at to wildlife and habitats preservation and constituted Indian Board for Wildlife. The Wildlife Institute of India (Dehradun) conducts research on the ecological, biological, socio-economic and managerial aspects of wildlife conservation. The SA centre for

Ornithology and Natural History also conducts research and extension activities relating to different aspects of natural history and ornithology.

At an individual level it was seen a marked rise in environmental consciousness, green consumerism and association with the natural conservation groups. The need for nature conservation is of course, growing but it is unfortunate that even after independence of the country, ancient woodlands are being destroyed, coastal marshes have been drained, birds are disappearing, pieces of grasslands have been lost, and row of bushes are grubbing out. This unprecedented destruction is set to continue as demands from housing, transport, industrial developments, farming, energy, extractive industries and services conflict with natural environment which is becoming less or smaller. To resolve the development versus conservation conflict is the primary goal of the principal aim of the local authority.

ROLE OF THE LOCAL AUTHORITY

The political will and supporting tendency of the people for nature conservation has resulted in regulations and legislation with the aim of conserving and improving the natural resources and environment. In response to their responsibility for nature conservation, the local authorities must have plans and strategies that include their objectives and actions. An environment auditing is not a replacement for these initiatives. It should seek to reinforce the authority's role, monitor progress on strategies, coordinate conservation with other policies. Where such strategy and policy is absent, it should be recommended to undertake these practices.

Different authorities, as discussed earlier, have used the approaches to environment auditing. The successive chapter will discuss on the differences of regulatory or external environment auditing and internal auditing. These two approaches are not mutually exclusive. The range of options combines the two.

STATE OF THE ENVIRONMENT AUDITING

While conducting environment auditing, the auditor's first consideration will be benefits of local authority by auditing environment. It is necessarily to be clear about the gain of the local authority from the very beginning because of considerable resources outlay. In regard to nature conservation, the main purpose of auditing is to get picture of the natural resources and to identify the changes or changes required and also to find out the way on which it can be managed further in a positive way. By drawing the information about the status of species and habitats as far as possible and making the issues clear and comparable, the local authority promotes decision making process and analysis. And, may also disseminate an information and develop educational resource for others. The local authority can monitor the trends over the time by updating the information and education resources.

When the authority affirms the need of environment auditing, the next step will be to define the scope of natural environment and the data level required for the auditing. The survey area will be certainly as of jurisdiction of the local authority's administration. While assessing the conservation value of particular land use types, this will certainly be the same. However, when dealing with wildlife species, the survey area may be considered in a wide range of the subject. For example, a shifting birds may spend the winter season in the jurisdiction of the local authority and then over summer in the overseas countries. In this case, the protection of these species in question may not be simply a local authority's responsibility. It requires consideration of global attention.

The data requirements for environment auditing are usually taken from secondary sources and involve a very little or no primary survey. In the field of natural conservation there is a chance of information gaps that need to be filled. But this should be a recommendation of an environment audit and it should not be made requirement of it. This is important because an adequate environment auditing can be drawn up with limited resources directed towards data collection and analysis. The sources of data may be the government departments, institutes on natural resource and wildlife such as WII, SACON, NNRMS, ENVIS centres, and/or Environmental NGO's.

On identification of the sources of data the next step is to collect the data and analyse it. In fact, the nature conservation is meant for protecting the habitats. In the process, separation of the natural environment into land use types and deal with each one will be the practical and convenient approach. The matrix elucidates the land use types, which can be divided into three categories i.e. Status-trends and Actions, while addressing the environment audit.

The following section deals with some of the issues involved in the three different land uses.

AGRICULTURE

Agricultural land will likely to be the largest single land use in rural areas. The land is classified by the national grading system within the Ministry of Agriculture. Topography, soil and climate conditions are the main determinants of quality. Information on the extent and distribution of each grade of land should be sought with details like nutrient content, erosion, fertiliser and pesticide use, micro-climate etc. The main sources of data are the annual statistics collected by MoA or other apex departments or subsidiaries within the ministry. The data provides information on the type of practiced farming, farm size, land use, livestock, ownership and employment. Some more information may be available which can be collected.

TREND

Agricultural trends may be assessed from structural factor and an environmental position. Structural factors contain employment, land values, crop area and yield, livestock operations and institutional arrangements. A balance is also being sought between agriculture and environment by enacting agricultural act/ regulation. These changes in agriculture policy have caused to move this trend. There is now more support for conservation initiatives such as agricultural diversification scheme, farm conservation scheme, farm woodland scheme and in environmentally sensitive areas. The follow and success rates of each scheme should be monitored. Because there is no planning control over agricultural land uses, an authority can have limited influence. So, there is a need of involvement of agencies with environmental protection and its improvement activities.

The important factors for assessing local trends are :

- land use changes
- effects on water, air,

- effects on foodstuffs,
- amenity and life quality
- effects on other eco-systems

WOODLANDS

Commercial fruit plants and shedding woodlands reflects quite different eco-system types and as such should be treated differently in an state of environment. Woodlands have a very important nature conservation role to play. They provide a habitat for numerous native species of flora and fauna, help prevent erosion and water logging, act as a windbreak for crops and settlements, help fix atmospheric carbon dioxide, are a useful amenity resource and are also a source of income.

The woodlands are valuable because of the stability and diversity of their plant and animal communities. These are widely distributed and rich species.

Data on possible extent, distribution and condition of woodland habitats and on species is available with the MoE&F, Government of India and may also be available within the local authority and the local conservation groups as well.

TRENDS

The species rely on woodlands for their survival. Their disappearance may result in a growing list of endangered plants, species and animals. Much woodlands cover has been lost due to over exploitation, removal for other uses, disease, air pollution and poor management. However, woodland habitats can be protected in the ranges, which may include national and local nature reserves, tree preservation, sites of specific scientific interest etc.

NAEB within the Ministry of Environment and Forests, Govt. of India provides funds and support for planting and plant management schemes. With regard to agricultural land uses, the factors impacting on the health of woodlands should also be evaluated. In case the information is scarce, the local authority may need to consider conducting additional research. Many NGO's and environmental consultants, now provide these services.

URBAN NATURAL HABITATS PARK

Habitats within urban areas or on the urban fringes are an extremely important wildlife reserve. Many animal species have found taking shelter in towns and cities and taking advantage of the supply of food from waste dumps, gardens and parks. Animal species such as foxes, rabbits, mice and rats and birds like the sparrow, pigeon, crow have become especially arrive at urban living.

The conservation value and potential of parklands should be surveyed and the wildlife quality be assessed. The data can be collected from the urban authority and the local conservation groups.

TRENDS

Urban parklands are designated as open place for public in development plans and, are afforded almost complete protection. The primary issue is therefore the type of management practices, which are undertaken to maintain the parks. Farms are kept as formal spaces with flowerbeds put into a good order, neatly grass by mowing and small lovely trees. There is an evidence of creating more informal areas within parks where wildlife must be given equal weight alongside the more traditional considerations of recreation and aesthetics. These areas are, in fact, easier to maintain, even cheaper to manage. Another important issue is that of access. Areas of relatively low wildlife significance may prove excellent from a recreational or educational perspective because of their location within population centres. In such cases their potential lies in raising awareness of nature conservation issues.

ISSUES INTO ACTIONS

On assessment of the land use resources, the authority may then in a position to present an overview of nature conservation in the respective region. This overview is very valuable not only as a summary for the policy makers and the public but also because of the inter relationships between elements of the living and non-living environment. The linkages in ecosystems are often poorly understood and therefore, it should form a distinct area of analysis.

The goal having priority over the environment auditing is the natural environment; therefore the result of the work must be an action programme to be implemented. These programmes should be defined in terms of locally sustainable environmental capacities. The issue of environmental capacity has already been discussed earlier. So far nature conservation is concerned, it involves a set of targets to be identified and a range of indicators for measuring the quality of environment. The targets may be set on the top by strictly applying the guidelines as outlined in the relevant policy and legislation. There could be an overall target to be in a position to provide statutory protection to all sites of nature conservation importance. The indicators measure the designated land area and the quality of wildlife in terms of species variety, its lot or rareness. All capacities have a standard definition and it is the necessary role of audit to identify the appropriate and reachable capacity. In this process it is to be underlined that the updated standards are maintained and the targets are continually revised and upgraded, where possible or required. It will facilitate the monitoring of environmental indicators and allow progress to be checked regularly.

The Checklists are

- Whether the land use types have been surveyed and its significant assessed.
- Whether the status, trends and actions issues have been addressed.
- Whether environmental capacities have been defined in terms of target with favouring indicators.
- Whether impacts and trends regularly monitored for all wildlife resources.
- Whether opportunities for habitat creation and improvement addressed.
- Whether the list of all designated areas and specific sites are viewed.
- Whether the ecological linkages within and between habitats are considered.
- Whether the database of environmental information have been used.

- Whether the database was user friendly and open to the public.
- Whether plans for periodic environment auditing reviewed.
- Whether the possible sources of information have been utilised in preparing the audit reports.

INTERNAL AUDITING

Internal auditing has not only concern with the state of the environment auditing for providing information collected or data gathered but its prior concern will be the improvement of the natural environment through the policies and activities of the local authority. The scope of the Internal Auditing may therefore, defined by the points of contacts between the environmental topic areas, which were discussed earlier, and the authority's internal policies and practices. While observing the role of an authority towards nature conservation, it will be useful to examine the different functional areas involved in the internal auditing. These are:

- Development planning
- Development control

Land management

These areas may vary at different levels of local authority or government. The general outlines of the activities and opportunities are discussed here. With regard to the resulting area to the environment auditing, the outlines are the basic requirements that may require elaboration and expansion in terms of each particular authority.

Development Planning

Development plans set up the overall strategy for ensuring a balanced approach between competing demands of development and environmental conservation which includes nature conservation policy.

Structure plan policies set strategic objectives, which may be further, developed in plans. Local authorities are required then to invite conservation bodies to submit information for plans and to consult widely on draft proposals.

Development plan policies on nature conservation should be listed with their well defined purpose and examined success alongwith highlights on the gaps in area and amount covered. Statement of policy should be clear, meaningful and comprehensive, indicating their firm commitment. There should be policies on all sites of conservation significance and the management of such sites should also be evaluated opportunities for new sites creation and further designations should be assessed, with particular attention to new developments and land to be brought under cultivation or abandoned land. There should be thorough consultation on these aspects and the public should be encouraged to participate in all development plan activities. It is essential that the sites of nature conservation significance are covered by statutory plan policies; non-statutory plans do not have the same legal status. Expert advice should be sought where necessary to increase in-house knowledge. All sites of wildlife significance should be mapped by using, if possible, geographical information system. It is always better to be pro-active to deal direct policies and practice in a positive manner rather than waiting for damage to occur.

Development control

Through the planning system, local authorities may exercise reasonable control over developments which may effect nature conservation in a positive way. Where habitats are under damaging threat, planning consent may be refused on conservation grounds. Approval conditions and management agreements may be used towards conservation interests. All the developments be checked against the regulatory schedule.

In case a state of environment auditing is not carried out, the map of all sites of nature conservation should be utilised to proceed further.

It is statutory duty of the authority to protect all such designated areas and it will be obligatory to conduct environment impact assessment for assessing the developmental effects on environment. Green belts and open spaces, however not specifically designated for conservation purposes, are having a valuable role to play in order of conservation. Woodlands and individual trees can be protected for the purposes of future use and pleasantness by using the technique of tree preservation.

GLOBAL WARMING AND CLIMATE CHANGE

1. The level of greenhouse gases in the atmosphere is rising due to increased human metabolic activity, particularly the burning of fossil fuels, the loss of forest, and the increasing intensity of agriculture.
2. This is leading, so most scientists believe, to the very rapid (geologically speaking) warming of the Earth - at the rate of between 0.3 and 0.7 C per decade. Feedback loops in the global ecosystem are not fully understood and may moderate the warming, but are more likely to exacerbate it.
3. Global warming, in turn, is affecting climate, and will do so at a greater rate next century. Climate change in many parts of the world is also being caused by land use changes, especially deforestation. The combined effect is to increase climate instability and, in the tropics particularly, the risk of drought or violent storm.
4. Warming also leads to ocean expansion, and *could* trigger polar icemelting, causing sea level rise and flooding of low lying coasts.

The development control system provides an opportunity to protect the habitats and improve the conditions of nature conservation. The planning applications should also be monitored and their likely environmental impacts be assessed. Where adverse impacts identified, the development may be refused on conservation grounds. Where impacts are less rigorous and harsh, more information and in special circumstances, an EIA may be requested. Planning conditions and management agreements should be used in a greater way and not only to restrict damaging activities but also to encourage positive conservation initiatives such as creation of the site and tree planting. A clear guideline is necessary for all the sites planning permission. It is highly needed in particular, on or near the sensitive sites. Incremental developments should also be carefully monitored so as to avoid cumulative and indirect impacts.

The conservation measures should also be discussed with other local authorities, public sector organisations, industrial houses, voluntary organisations, and the public so as to gain clear advice and provide guidelines to encourage co-operation. Planning in general and design guidance on protection, management, and site creation for developers would be particularly helpful in getting the local authority's message across. Public open spaces should also be vigorously protected and managed, keeping in mind the possible conservation.

Land Management

In fact, local authority can use its policies and practices of nature conservation on its own land to set an example for the community. There are a lot of opportunities under which existing sites are preserved and enhanced, even newly created sites can be preserved and improved. It is responsibility of the local authority to manage the land in the light of existing nature conservation potential. Classification of sites considering the sites of conservation important, parks and graveyard, and the sites having potential to create new habitats of value to wildlife may be useful. Habitat type should be mapped and species surveys be carried out where they do not already exist.

Management practices of these different groups of sites may vary site to site. It requires comprehensive management strategy, which should cover a range of levels, from surroundings conservation strategy to site specific management plans. The former should be the components of an overview of the local body's policies towards conservation sites and the environment, overall aims and strategic guidance. The latter should focus on a particular habitat or habitats group in the specific area, management practices-objectives and the means by which it will be met.

The emphasis in all conservation strategies should be on enhancing the value of the wildlife sites. The formal policy guidelines - developed with the participation of various departments and outside bodies- provide a point of reference on which the progress can be evaluated. It also eases the consideration of nature consideration in the processes of decision making, development plan and the practices.

The followings are the issues which may be considered:

Fertilisers and Pesticides
Native species
Peat
Management
Education
Site creation
External sites
Pesticides and fertilisers

Generally, artificial chemicals should be avoided. It must be avoided to use on the sites of conservation importance. More organic methods should be applied such as composting leaves, grass cutting or outer skin of tree leaves.

Native species

Use of native species may be preferred to unusual varieties. They support more rich and variable ecology. Generally it is more adjustable. The wildlife value of species may be considered in new plantation and the management of existing area both.

Peat

Peat (carbonised vegetative materials) lineage/extraction is extremely destructive of natural habitats. Its use should be discontinued, as many alternatives are available.

Management

The maintenance of the site should be on regular basis and be carefully considered. The emphasis should be on low input and not-intensive practices. It is possible to improve wildlife value without undue expenses or it can be done by reducing labour and material cost.

Education

Public awareness is the key of effective nature conservation. Those who are responsible for land management should have a greater role to play in awaring the people in general. Sites should be promoted and explain the meaning of land management and conservation to the wider group of people. On site information panels, public guide walks or open days can be used to explain the purpose of a wildlife area. As a learning system, this information can be disseminated to the viewers own environment (e.g. school, home or work place) could raise awareness and influence behaviour in a positive manner. People's participation and support should be ensured at all the levels of activities.

Site creation

To bring the abandoned land under cultivation, to restore the refuse overbalance and mineral working, to improve ignored and less used land and new development areas offer opportunity to enrich wildlife quality by creating new habitats themselves and supporting others to do so.

Sites not under authorities control

Sites of importance may be secured through purchase or lease or by designating such sites as land for natural resource management and compensated thereof. A firm commitment should be made for conservation of the sites. A secondary form of safeguard the site is through making agreement about the use of land and its management with the owner. Such agreements can be made formal arrangements, memorandum of understanding or through enforceable management agreements with financial incentives.

Now we consider the resources available for nature conservation. These are

- resources within the authority
- resources within the community

Resources within the authority

Elected members

It is supposed to that the local government or authority is constituted with the elected members. In fact, the elected members are the great asset towards nature conservation, if they are aware of the high need of the nature conservation and they have strong commitment and involvement in nature conservation practices. It is only the elected members who take decisions on resource allocation, make policy and initiate for good practices. The standard measure of nature conservation commitment within the authority is either a committee or sub-committee with overall responsibility of environmental issues and nature conservation whether exists or not? The level of their commitment can be observed by reference of committee papers and statements on nature conservation and also by comparing budgetary allocations to conservation activities.

Professional staff

For successful nature conservation initiatives within the jurisdiction of authority, it is pre-requisite that the professional staff is having sufficient expertise to expose the nature conservation needs and to proceed in right direction. The authority is supposed to have at least one nature conservation officer with them. Comparatively large authorities may have an environment division capped with various skills on environment.

It is important in order to have an interdepartmental coordinated approach, to instruct them in an effective direction of resources and implementation of strategies, to have liaison / public relation officer for the purposes of liaisoning with other bodies and people concerned. The nature conservation officer-having expertise on environmental/nature conservation issues and professionally qualified should be posted at such a place from where he can be able to maintain wider contact-whether interdepartmental or public. The other staff may get skill for nature conservation by participating the governmental or non-governmental training schemes. All such opportunities can be availed.

Information

Information is an essential tool for development planning, development control, land management and also for formulating comprehensive conservation strategies. The purpose of an internal environment auditing is to collect all relevant information and make it available in an accessible format. If a local authority does not have sufficient in-house expertise in assessing the significance of a habitat or to formulate strategies, they should be able to call in expert advice.

RESOURCES WITHIN THE COMMUNITY

Local authority may be a strong player in the private and voluntary sector by encouraging positive conservative actions, in addition of auditing their own policies and practices on nature conservation. The authority should make themselves well aware of the activities of these (private or voluntary) bodies and provide financial assistance in the shape of grants and material aid to support them and their activities. A local authority can often usefully provide a coordinating facility to these groups for motivating peoples action and promoting collaborative work.

The aim of the local authority should not be to preserve the countryside but to conserve and improve it for the essential value and for the better quality of life. And, this can only be achieved by producing a more positive attitude towards nature conservation.

CONCLUSION

The Internal Audit - on completion - should make use of all natural resources towards improving policies and practices and thereby protecting and enhancing environmental quality. The state of environment auditing feeds into this, by providing baseline data and monitoring progress.

Determined attempt should be made to become aware of the areas of nature conservation importance and policies followed to afford a full range of protection.

Even sites of local importance only have a role to play in maintaining biodiversity. This is, in particular, true for wildlife passages such as rivers, canals, railways and boundaries as they have vital links with the habitats.

Protection policies need to be strongly supported by the decision makers of the authority with firm determination and commitment. The source of this commitment may be checked from the observed value of nature for humans, for sensitiveness to beauty or art, for recreation, as a large room for storing resources. This is, of course, a valid basis for proceed towards nature conservation. But, it should be backed up by recognition of the essential value of nature and the moral obligation to act as warden of the environment.

Checklists - Internal Auditing

Development planning

- Whether nature conservation exists in all the development plans.
- Whether the policies are well spelt, clear and comprehensive.
- Whether the development plan contains habitat enhancement, new site creation opportunity and designation of sites having wildlife importance.
- Whether the planning for nature conservation were made with consultation of the public.
- Whether the development plans include zoning atlas, lists and maps of all sites having conservation importance, described in details and such specific policies are outlined.

Development Control

- Whether the development proposals are checked against the regulatory schedule.
- Whether the development proposals are checked against zoning atlas, other lists and maps of the sites having nature conservation importance.
- Whether the authority requested for an EIA for all developments.
- Whether tested techniques applied for preserving the trees and woodlands.
- Whether the planning conditions are used to protect and promote conservation interests.
- Whether sufficient consultation and co-operation ensured with the participation of concerned parties.

Land Management

- Whether the authority's own land has been assessed fully in the light of nature conservation potential.
- Whether the authority is having formal wildlife conservation strategy.

- Whether there is a management plan for each site of importance.
- Whether the authority owned sites are managed keeping in mind the need of nature conservation.
- Whether the sites known as of wildlife importance are leased or purchased.
- Whether the formal or informal management agreements are entered into, with private landowners to strengthen environment friendly practices.

Resources

- Whether there is a committee with responsibility for nature conservation.
- Whether there is a full time conservation officer or wildlife officer.
- Whether there is a separate budget allocated for nature conservation activities.
- Whether there is a register of all useful skills with the authority.
- Whether expertise services undertaken to strengthen in-house expertise.
- Whether the powers been exercised for reducing air, land and water pollution.
- Whether grant-in-aid from government or governmental agencies received and exercised.
- Whether an environmental forum constituted or exists.
- Whether there is coordination between on nature conservation, recreation and education.
- Whether local conservation groups aided.
- Whether other assistance provided to local conservation groups - in kinds or financial.
- Whether the opportunities are taken to motivate community action and promote collaborative arrangements.

CHAPTER 5

ENERGY

About half of India's total energy consumption attributes cooking energy. No other form of energy use has greater impact on the environment. It is very important for human survival. Firewood, cowdung, bushes and crop wastes-falls under the category of non-commercial energy-provide more or less 80 per cent of the cooking energy requirement. Not only in the villages but as much as 40 per cent energy consumed in the towns/cities from non-commercial sources. It is important to study the pollution aspects of non-commercial energy. More or less 110 million tones fuel wood, 60 million tones cattle dung and 30 million - tone agriculture waste consumed in household sector. Though we are using this source of energy for generations and we have liking to use them till today. We are not aware how much we do pollute the atmosphere and ecology from these sources. In fact, we are releasing or expected to release million tones of pollutants to the environment from non-commercial energy sources.

WHO's report identified four definite links between poverty and energy/power supplies. The links are:

- A rise in mortality
- Respiratory disorder in children
- Depression amongst women at home
- Owing money in low income households

While excessive waste of energy makes worse social problems, a conservation approach to energy conservation can certainly reduce such problems. General or specific health problems are directly related with the consumption of energy in households.

There is an evidence of seasonal depression to some degree due to shortage of grow light. Asthma /arthritis in children is often linked to abridge size of the home-crowded together/closely compacted.

It is a good sign that the government has managed to reduce indoor air pollution from household fuel consumption on a large scale, largely by expansion of gaseous fuel supply, and, there are good reasons to believe that serious indoor pollution will disappear in the next twenty years as living conditions continue to improve. But at the same time, this thought, and the fact that most households in cities are not effected by indoor flue gas, may have given policy makers the illusion that the indoor air pollution from fuel burning is no longer a public health threat.

Such optimism would be premature, since there are still hundreds of millions of people using solid fuels for cooking and/ or heating, and the population exposed to serious indoor air pollution may be similarly large. In fact, the lack of updated research and extensive monitoring data conceals the scale and severity of indoor air pollution, preventing the development of an alleviation strategy.

The first order of business should be to carry out a national survey of the status and nature of the problems so that a clear policy agenda can be put together. This could be done one of the areas where immediate actions and continual efforts can have a thorough impact on public health, since those who are exposed to high level indoor air pollution are most likely to be poor urban and rural households. Controlling indoor air pollution has the double advantage of improving health and bringing environmental equity to the poor and less fortunate.

ENERGY PLANNING

By establishing a strategic plan and rational systematic approach, a sizable quantity of fossil fuel/energy can be reduced. The auditing process may assess the progress in terms of energy conservation achieved by the local authority. The progress path can be determined as follows:

Whether the authority has a well-defined environment and energy policy - Collective commitment towards energy conservation is the first and foremost step. This commitment should be achieved through the authority by adopting an energy and environment policy in a wider range. This policy provides a positive direction and sense of commitment (in action also) on behalf of the communities functionaries.

Whether the energy plan is prepared - it contains analysis of the current situation in generation and use, and plan for energy supply irrespective of non-commercial, commercial, renewable, combined heat and power category, programme for improving energy related to major sector of use. It will set out a strategic plan of action. Energy conservation will be a major programme in this action plan.

Whether it has established current energy programme - it is imperative to establish a baseline from which one can measure the future energy consumption behaviour. Macro energy consumption can be measured by the power supplied for fuel utility such as gas, electricity, heating oil, transport fuels, coal etc. - supplied to the area. Power generated from manufacturing company or privately may be taken into account. This total consumption has to be attributed to the different sectors such as private or public housing, commercial or public buildings, industry, transport, agriculture sector, lighting in a public place etc. An estimate is made of approximate conversion efficiency in each sector.

Whether it has monitoring system - the best method of monitoring is to set up a database of the building stock and energy use. A useful subset of this exercise may be the energy labeling of the existing housing stock for which the local authority is responsible. This may be an effective tool for measuring the total environmental impact of households on the environment.

Whether it has adopted a set of specific measures - for example, to ensure future planning applications for commercial buildings to include an environmental assessment, or, to ascertain that all government buildings to go for an environmental assessment, to plan upgrading of all public housing, Incentives or controlling measure to encourage combined heat and power schemes, encouragement to set up a local energy resource centre to improve

Table 5.1 : Extract from energy section of audit report

Environmental aim : improve energy conservation in council buildings and encourage others to be aware of the need for increased conservation of energy

Issues	*Existing Policies*	*Actions/programmes*	*Appraisal*	*Recommendations*	*Agreed Actions*
Reduction of energy consumption	Council housing stock : Housing service Aims to invest continually in the Council's housing stock to preserve and improve its condition	Affordable Warmth Programme : More comfortable tenants Lower fuel bills Fewer condensation problems Reduced CO_2 emissions Saving energy and fossil fuels Good value for money Includes some promotion of Affordable Warmth aims to private sector through exhibition meetings and competition	Structured documented programme with agreed policies is being implemented Monitoring the programme's performance could be improved No long term PR and awareness strategy prepared	Methods of more effective long term monitoring of the achievements of the Affordable Warmth programme should be investigated and implemented Long-terms PR strategy promoting MDC's role in a national flagship scheme should be prepared	Investigating monitoring of individual tenants Sheltered housing communal areas energy consumption to be investigated with Senior Housing Officers and Tenancy Senior Officer Date from sheltered housing schemes to be included in current office monitoring system Long term PR strategy for Affordable Warmth Programme to be included in corporate PR strategy

local energy conservation measures, to conduct training for the managers of public houses.

The local authority should follow the energy planning measures, as follows;

formulating energy plan
support for establishing energy centres
assigning a green surveyor attached with housing department
assigning a manager for energy attached with engineering department
energy awareness training to the managers
establishing energy and environment plan
energy labeling initiatives of public houses

ENERGY MANAGEMENT OF THE AUTHORITIES OWN BUILDINGS

On priority basis, the energy issues should be taken and tackled. To begin with, the mission statement, which will include organisation's commitment to energy conservation, should be written. The energy management declaration may be in the style of responsible energy management and improving energy efficiency is commitment of our company. And, it will reflect through the following actions;

- Publication of a corporate energy policy
- Establishment of an energy management responsibility structure
- Meaningful awareness of energy efficiency among employees
- Establishing regular review system
- Setting up of improvement and performance targets
- Monitoring and evaluation of performance levels
- Provision of report about performances and improvements to employees and the shareholders

Assigning an energy officer

The next step will be to assign an energy officer with the responsibility for carrying out the commitments of the organisation and on behalf of the organisation. The job responsibility for the energy officer could include a target to be achieved within the set time frame. The target may be fixed to reduce 10 to 15 per cent energy uses in a year. On establishing the current situation, the officer may check whether the organisation is on the advantageous tariff structure with the fuel utilities or not. In a commercial organisation, often one finds a serious mismatch, which may be costliest.

The next step will be to identify all the plant and equipment that consume energy and also the extent of consumption. Once it is established where the energy is being consumed, the decisions based on reason can be taken for minimising the wastage. This may require priority of "do it now" through the issues that of "do it in the financial year" such as replacement of lighting and heating controls, replacement of inefficient plant i.e. boilers or switch to a different fuel. All this can be written into a strategic plan for approval of board and a budget allocation.

In case, there is air conditioning then the heat from inefficient lighting will add to its load.

IMPROVING AWARENESS LEVEL OF THE HUMAN RESOURCE

While approaches made for reducing the waste of energy, changing behaviour in terms of the use of energy and changing technology will be effective. The mission statement on energy may be a part of wider document on the intentions of the organisation regarding energy and other environmental issues, which should be distributed to all working staff or widely publicised, asking for cooperation of all concerned in terms of energy saving and other conservation issues. And, it may be hoped to get positive response.

AN AUDIT OF HOUSING STOCK

An audit of housing stock switches on the authority to establish a line marking position of the energy performance and conditions of its properties. It provides information on the distribution of certain property characteristics. It is the first step to assess the efficiency benefits of annual maintenance improvements to the stock. The improvement programme also is assessed. This can be produced in average energy rating for the stock and pollutants emissions. A baseline study also provides a corrective measure and necessary pre-conditions to a housing energy and environmental policy thereof.

Followings are the key steps for a successful energy audit:

- Assign a staff member to coordinate the audit on the stock and to expose the results
- Collect information on tenant's circumstances and assess their needs. Arranging meetings of the tenants group or representatives, information and views can be collected.

Identify existing sources of information for the age of the houses, built shape, heating and insulation in the stock. These can be collected from property registers, existing stock database, stock survey's report, building control records, installation records, servicing contracts or the maps.

- Decision on measuring the energy efficiency of the stock, computerised stock profile, average energy rating and pollutants emissions can be assessed before selecting a system of rating while on audit.
- Decision on the rating system selected in terms of level of rating requirement for the stock, precise rating of each property, if required and processing of information. It should be ensured whether the information is a part of an energy policy.
- Decision on the key indicators for the stock in terms of producing stock profiles of energy rating and pollutant emissions.
- Decision on updating of the stock profile and its periodicity. Selection of staff
- Identification of energy related work and investments
- Dissemination of information to all relevant staff.

ASSESSMENT OF ENERGY PERFORMANCE OF THE HOUSING STOCK

The existing tools will produce an assessment of the energy performance of the stock -an energy rating. It may be based on standard assessment procedure or may use a number of parameters but incorporates standard assessment procedure as a part of the rating. Carrying out a full audit of each individual house produces an energy rating. In most of the cases this is, however, likely to be extremely time consuming and expensive. The method can be simplified by carrying out an audit of houses type and then calculate the results from known data for particular type of houses. Applying the number of parameters, which may require about each house, can produce different levels of rating. A low level rating based on half the questions of high level rating will produce only an approximate energy rating for an individual house, but for the stock as a whole becomes highly accurate.

INFORMATION SOURCES AND UPDATING

It will be useful if sources are listed for all the information, which may be considered to make up the audit of the housing stock. If it is stored in a database, consideration must be drawn to the interface with other storage systems and planned development. To get real benefit, the stock profile needs to be updated regularly, at least once in a year. This practice helps towards energy efficiency objects and targets to be marked or tracked. In the coming years, the accuracy of the rating can be improved by feeding in additional sources of information, options appraisal, maintenance visits and tenants comments. It is planned to improve the accuracy of the rating. This should be kept in mind while deciding the system to adopt.

REHABILITATION

The internal energy audit may extend outside the scope of assessing management and monitoring procedures. Rehabilitating an old building rather than demolition and replacement can itself be an energy conservation measure. A sizable proportion of the energy budget of a building may be in the energy content of the materials in its wall, floor and roof.

Much can be done to upgrade the energy efficiency or reduce load of energy in the old buildings even. The building can be given to new outlook/direction in energy saving direction and by projecting window of an upper storey. By this way glazing to the walls can be reduced. Often it is observed that there is a capacity for the sun spaces, particularly in the neighbouring buildings, and also the provision of light wells to increase day lighting in dark area. The building can be protected thermally by imposing non-conductors to the gallery, hall or upper room, building hollows, walls and floor. If the externally or internally surface need replacement, some insulation measures may be the only economically viable. If the surface treatments have to be done anyway, the extra cost of adding the insulation is minimal. If ground floors have to be re-laid, that they should be insulated, either by unbendable slab applied to the surface level, a coverlet between the supporting beam in a floor or an insulated flooring board.

Any replacement windows should be double glazed with a sufficient air gap and filled by inert gaseous element. Windows retained should be secondary glazed.

In case, renewing heating system, at the first sight it is to be considered a change of fuel from used electricity to natural gas. Condensing boiler may be installed. If retaining the existing heating system pump overrun and weather compensation is considered. Thermostatic radiator valves may be fitted.

If replacing lighting then provision should be made for fitting of electronic fluorescent lamps for corridors passages stairwells, security and external lighting.

If the building is overheating in summer then solar control film fitting may be considered. This will reduce the load on the ventilators or air conditioning system.

If hot water consumption is huge in the building then fitting of solar collectors be ascertained on priority basis. In commercial or public buildings, occupancy control measure is considered, especially in terms of lavatory blocks.

NEW BUILT

One of the areas of greatest potential for energy conservation in the design of new buildings is to maximize for the passive use of solar energy, both in healing the building directly and also in creating passive stack or vortex ventilation without the need for any mechanical assistance. The audit should assess the degree to which new buildings built or sanctioned by the authority are recognizing such techniques.

Unless the building has exceptional demands (such as a lecture hall or financial dealing room with a very high cooling load from all electronic equipment) there should no need for air conditioning in the U.K. If a new building designed for an UK site incorporates air conditioning then one has got the wrong building (or the wing architect)

Installing a wet under floor radiator system can be a move towards a low energy building and increasing health and comfort level. Wet under floor systems are particularly well matched to condensing boilers because the heating circuit is run at only about 35°C. Using a wet under floor system can save about 20 per cent of fuel over a conventional radiator system. Such systems are installed in 55 per cent of domestic housing in Switzerland and Germany and 50 per cent of dwellings in Scandinavia but there appears to be a taboo against the use of such systems in the building industry in the UK.

The Energy Technology Support Unit at Harwell has monitored private passive solar houses built in southern England over the last five years. These have regularly turned in space heating bills of 10-15 a year and the technology have already moved on since these housed were built. In 1991 there were 40 houses built in Europe requiring a zero heat load.

While the perception of the house building trade has often been that unconventional solar houses have sold readily, and some have attracted a premium on the price.

The Future

Provided the house is of a passive solar design, It is now technically possible to build domestic housing in the UK that does not require connection to mains energy. Electricity for lighting and appliances can be provided by small-scale photovoltaic (solar cell) systems either alone or in conjunction with small wind or water turbines. Water heating can be provided predominantly by solar collectors with back up from the independent electric system as a dump load. Because of the increasingly high cost of connection to mains services. If one new building on a green field site in a rural area. This will be the most economically favorable option by the year 2000. The attitude of local planners is likely to be the biggest hurdle.

ENERGY MANAGEMENT DECLARATION OF COMMITMENT

This company is committed to responsible energy management and will promote energy efficiency throughout its operations by the following actions :

- Publish a corporate policy
- Establish an energy management responsibility structure
- Increase awareness of energy efficiency among employees
- Hold regular reviews
- Set performance improvement targets
- Monitor and evaluate performance levels
- Report performance changes and improvements to employees and shareholders.

Other options are likely to present themselves over the next few years. Which go beyond the conventional. Local authorities should be open to change in building practices. The high cost of new housing and the issue of homelessness may well lead to the introduction of cheap self-build timber frame kit housing. Similarly while the average size of households, especially in the cities. Is getting ever smaller, there is likely to be another movement towards people living in small intentional communities already receive government subsidies.

The movement of middle class families from London to rural Devin has prompted Devin Country Council to set up a special unit to study their new needs. The incomes to rural areas are unlikely to want to import suburbia with them and the local population certainly will not. One response to demand for new rural dwellings is likely in the next twenty years to be a new architectural movement for hidden housing This movement will be characterised by designing building that blend into their backgrounds so that they can only be identified when one is close up to them. In this case such elements as earth sheltering. Turf roofs, rammed earth and green timber building and reflective glazing will all be in demand.

What will be the attitude of the next generation of planners to such new ideas? Generally politicians and officials tend to reflect what they perceive to be public opinion (which in reality has already moved on ahead of them.)

CONCLUSION

Reviewing this chapter new should give the reader as sense of how much could be achieved by a national commitment to energy conservation. The Building Research Establishment believes that simply by utilising only those technologies and techniques that is currently cost effective. The energy consumption of UK building could be reduced by 25 percent. If the best currently available technology were used then this figure would rise to 50 percent. This is a considerable challenge but it should not be viewed as one with only costs attached. There are many positive benefits to be gained by meeting the challenge. With hindsight it could be argued that western industrial society tends to solve problems by creating two more. One of the indicators of an environmental approach to problem solving is that tackling one problem also helps the solution of two or more others.

Thus a concerted effort by national and local government by developers and the building industry, by managers and homeowners would have many secondary benefits beyond saving energy (see box 6.4). The energy audit can be used to trigger this effort on the part of local authorities.

The national benefits of saving energy

- Reduction of CO_2 emissions which contribute to the greenhouse effect
- Reduction of NO_x and SO_2 emissions which contribute to acid rain
- Reduction of dependence on ozone depleting refrigerants
- Reduction of environmental pollution from fossil fuel exploration
- Increasing of people's real income and company profits and a reduction in tax burden.
- Reduction of social problems associated with fuel poverty.
- Revitalisation of depressed construction industry.
- Encouragement of renewable energy industry and reduced dependence on fossil fuel and nuclear power.

- Buffering of economy from worldwide fluctuations in primary fuel prices.

Checklist

Energy management

- Whether there is a mission statement on energy?
- Whether an energy officer has been appointed?
- Whether a monitoring programme has been instigated?
- Whether realistic conservation targets been set?
- Whether fuel tariff structures are appropriate?
- Whether a current energy use audit has been undertaken?
- Whether a plan has been made set to achieve target?
- Whether an awareness programme has been launched?
- Whether the performance of the energy officer monitored?

HOUSING AUDITS

- Whether sufficient information on tenants heating circumstances and needs has been collected?
- Whether the current budget for energy related to housing improvement and area of expenses identified?
- Whether information on the condition, design and insulation of the dwellings has been assembled and arranged systematically?
- Whether energy efficiency rating method of housing stock considered and requisite information/needs included in the internal audit or not?
- Whether the key indicators, used in the audit identified in terms of average energy rating and pollutants emissions?
- Whether frequencies of updating the stock profile considered, and staff input and information source identified?
- Whether the recommendations based on the findings of the audit has been conveyed or being conveyed to the staff concerned?

REHABILITATION OF OLD BUILDINGS

- Whether reorientation of the building has been considered?
- Whether sun spaces or light wells can be added on?
- Whether cold bridging points have been covered or insulated?
- Whether extra insulation to extra container has been considered?

- Whether windows and doors have been drought proofed?
- Whether double or secondary glazing has been evaluated?
- Whether benefits of the fuel change ascertained?
- Whether upgrading of heating controls considered?
- Whether low energy lights have been installed?
- Whether occupancy controls have been considered?

DESIGN OF NEW BUILDINGS

- Whether design has been maximised for passive solar heating?
- Whether submissive ventilation designs have been considered?
- Whether heat recovery ventilation system has been applied in the whole house- being constructed?
- Whether design maximises daylights?
- Whether there is a provision of super insulated covers?
- Whether low emissivity glass used in the double-glazing?
- Whether the heating system considered the object from which heat radiates?
- Whether the wet underflow radiator system has been considered?
- Whether condensing boilers have been fitted?

COMMERCIAL BUILDINGS

- Whether the circuits of light and heat have a well-defined region?
- Whether the heat pumps have been installed?
- Whether micro climatic heats and powers unit has been evaluated?

LONG TERM POTENTIAL FOR THE AUTHORITIES TO CONSIDER

- Use of timber frame, kit housing
- Earth sheltering, turf roofs, rammed earth, green timber.
- Glass houses.
- Solar power, wind power, water power, mini CHP
- Heat stores.
- Water recycling.
- On site biological sewerage treatment.

CHAPTER 6

TRANSPORT

In fact, An environment audit of transport policy may be concerned with the implications or level of accessibility, and not directly concerned with measuring accessibility. So far accessibility is concerned, it relates to basic questions of individual choice, economic growth and equity, therefore while raising the environmental issues profile, the audit may challenge the set goal of public affairs. It may redefine or reinforce the goals set by the political desire. By this standard, environment auditing of transport is subject to cause an argument.

However, motor vehicle emissions have become major source of ambient air pollution in a few large cities, which includes Delhi (NCR), Mumbai, Kolkata, Chennai, Kanpur. On average, mobile sources account for more or less 50% emissions of NOx and about 85% of CO emissions in the cities. It is presumed that the urban air emissions caused by urban transportation growth will be the main sources of urban air pollution. The fast growing fleet of diesel driven trucks and buses in our country should also be a matter of serious concern for air pollution regulators. Soot (black powdery) emissions from diesel engines are a serious health concern due to their substance, which produces cancer. The government has already developed a national strategy for reducing motor vehicle emissions based on the experiences of global situation. Key components include phasing out leaded gasoline, tightening emission standards for all categories of new emissions, adoption of cleaner fuels, upgrading vehicle inspection and maintenance programmes and implementing traffic and demand management. The critical issue is, now, to implement the strategy in an effective manner, strictly in a time frame.

Early signs have been encouraging because of the strong push of the Supreme Court of law to implement the national strategy on transport. Kolkata, Mumbai, Delhi and Madras are the cities that are effected with the serious motor vehicle pollution in the country, are moving aggressively to control motor vehicle emissions.

The country is getting to the stage here a significant number of urban residents are able and willing to purchase cars. Multinational and domestic manufacturers are also maneuvering for position to market affordable cars to private citizens. As a matter of concept, a tightly planned city with highly efficient public transportation system would serve the public and the environment as well, but realising that would require come together with the multiple government policies and market forces. As general principles for reducing urban congestion and motor vehicle pollution, large cities should try hard to apply mass transit systems to the primary urban people movers; and should make a plan and develop in patterns that can be most efficient use of mass trust systems and also implement policies that discourage private car use for commuting to and from work. A division of labour and responsibility between national and local agencies concerned with vehicular emissions control is desirable. In broad terms, the national agencies should focus on managing manufacturers of motor vehicles and fuels through:

- by structuring emission standards for all categories of motor vehicles and ensuring that manufacturers are complying
- by structuring quality standards for fuels and ensuring that refineries and importers are complying
- by determining vehicle fuel efficiency guidelines and promoting compliance, and encouraging research and development of cleaner and more efficient motor vehicle technologies
- by building institutional capacity for standards enforcement, including clarification of administrative responsibilities of concerned national agencies and local agencies, provision of financial support, and establishment of testing and inspecting procedures and facilities

- by assisting provincial and local body governments to build capacity for implementation of national and local standards

LOCAL AGENCIES SHOULD FOCUS ON

Enforcing emission standards for operating vehicles through normal vehicle registration procedures that require periodic inspection and maintenance of motor vehicles monitoring local air pollution and using it as guidance for plan measures to control motor vehicle emissions, using options such as traffic management through regulation and /or economic incentives planning for and developing more efficient urban transportation systems, and working with the national government on appropriate financial incentives to support such efforts.

This chapter discusses the maters related to transport audit in six sections:

- transport trends and environmental issues,
- tackling the issues of environment in terms of transport,
- choice of indicators and targets,
- evaluating transport policies,
- problems of achieving policy firmness,
- review of internal practices.

TRANSPORT TRENDS AND ENVIRONMENTAL IMPACTS

While most other sectors of energy use have shown marginal increases, the transport sector shares risen. Within the land passenger the total motor car dominates. The fuel use may be more than 80% of the total fuel used. The increase in passenger transport energy use reflects long running social and economic trends. It is not only that more people now own and use a car but also they make more trips, and each trip is on average longer.

Pollution is treated by the transport system as largely uncosted externalities. Transport is heavily implicated in a range of pollution problems that include smoke-laden fog, acid rain and global warming. Transport accounts CO_2, which cause global warming. Transport energy use also adds to emissions of other key "green house" gases such as nitrous oxides and low level ozone. If we wish to halt global warming, we have to cut the CO2 emissions in a sizable quantity. International agencies and the national government recognise the need of urgent action to stabilise global ecology.

Environmental impacts of transport

- *Global pollution* - emission of gases CO_2, NO_2, and indirectly O_3 : Contribution to acid rain SO_2 and NO_x
- *Local pollution* - health and fertility effects of carbon monoxides, effects of lead on child mental health, effects of sulphur dioxide, black smoke, volatile organic compounds, and low level ozone.
- *Resource use* - increasing use of the limited resources of oil, use of metal and other non-renewables in manufacturing, use of scarce land resources for roads and parkings.
- *Aesthetic impact* - visual impacts of roads on town and country as well.
- *Physical impact* - apprehension of pedestrian and cyclists, accidents, creation for barriers to pedestrian movement freely.

TARGETS FOR GLOBAL SUSTAINABILITY

The environmental audit needs to identify clear and sustainable targets against which to measure progress in the local area and access policies. The level of CO2 emissions can be taken as one key indicator. The government has shown their commitments to stabilising emissions, however, it is very difficult to achieve stabilising targets and it requires reversal of long term trends. It also requires concerted action on all fronts by the relevant agencies:

- effective and consistent transport strategies,
- land use planning aimed at reducing the need to travel,
- the substitution of telecommunications for travel,
- technological innovation in engine and vehicle design,
- increased costs for motoring (through fuel tax and road pricing)
- local authorities can have a significant impact on only the first two of them. But they can also try to influence other agencies and the public in general, by example, in their role as consumers, employers and operators.

REVIEW OF INTERNAL PRACTICES

The problem of unsustainable transport practices may be as much one of culture as it is application of science or finance monetary support for an enterprise. Habits of reliance on car should be extremely avoided. So as with other aspects of corporate behaviour, a prime purpose of any innovation to change the attitudes of employees generally and key decision makers in particular.

The main elements of in-house transport practice, all of which should be monitored and assessed on a regular basis, are:

- the choice of vehicles purchased,
- the quality of vehicle maintenance,
- the use of vehicles in providing public services,
- travel by staff for work purposes,
- and, the staff tour to work.

These five aspects are examined, and the key issues summarised in the checklist in succeeding paras.

CHOICE OF VEHICLES

The choice of vehicles for purchase by the authority can make an impact on the environment for a numbers of years. Two key decisions are the size of vehicles and their fuel. These decisions affect budget as well and it is understood that economy and environment friendliness can run in close association to each other. Where there is a choice between petrol, diesel, electric or CNG vehicles the arguments on impact are may be of doubtful or double meaning. In most situations the greater fuel efficiency of diesel motors should influence the decision, but in areas of high air pollution more detailed analysis of comparative impact is appropriate. Some authority may encourage use of natural gas or biofuels through their purchasing policy.

MAINTENANCE

Better maintenance method are probably as important as the choice of vehicle. Maximising fuel efficiency and the minimising of emissions. Maintenance contracts can specify performance standards.

PUBLIC SERVICE VEHICLES

It is more difficult to assess whether the vehicles are used for public services, that are provided with the minimum necessary number and length of motorised trips. Such assessment has to rely on the judgement of managing executives, the role of the auditor, therefore, is to observe whether the various service departments have given sufficient weight to the energy use in their decision process, and they were well aware of the fuel efficiency and gaseous emissions.

An essential part of raising awareness is knowledge. If decision makers have access to regular information on;

- number of trips made, by purpose and by mode
- level of emissions implied by the mileage,
- amount of fuel (diesel or petrol) purchased.

Then they are much more likely to make sensible decisions. All service department should be involved, not only those which make regular site inspection and client visits. Two of the largest energy users are the refuse collection service and the parks or open space maintenance services.

STAFF TRAVEL ALLOWANCES

The need to use a vehicle for a particular trip may be a possible action point. Individual car allowances may encourage excessive cost. Considering the fuel efficiency and less emissions, the car sharing or using public transport or bike should be promoted. Encouraging multi-purpose trip can be valuable- one trip serving several different purposes rather than several trips. It can also be substituted by phone, fax, computer link-ups to the possible extent.

TRAVEL TO WORK

The location of offices and other worksites is critical to mode of transport chosen and length of journey both, compelling the use of public transport, foot or pedal by employees. Within the constraints of any given site the authority can make things easy for their environmentaliy and cost-conscious workers:

- easy and direct access routes for pedestrians and cyclists
- secure bike parking, and cyclist showering provision,

- bus passes/bike loans be provided instead of casual car user allowance,
- limits on car parking provision, with regulations benefiting car shares,
- car pools at work instead of car allowances, allowing employees the option not to use or own their own vehicle.

STATE OF THE ENVIRONMENT MONITORING

It is very difficult to convert broad emission targets into practical criteria for evaluating current trends and policies. Local figures on energy/fuel use and emissions are not yet available in details and as required. The state of environment monitoring will rely on standard measures of transport activity and environmental impact. Information on trends in car ownership and use, public transport users, congestion or delay levels, accidents and aspects of noise and pollution to some extent.

The key indicator is the total amount of traffic from the sustainability angle. Total traffic level are correlated strongly with energy use and emissions. Traffic monitoring is more concerned with specific areas centrally located than the overall. It can be misleading, if we rely on such limited measures. However, since it is possible for inner urban traffic to be effectively restrained while at the same time the growth rate of traffic in the outer suburbs and rural district beyond a river or coast is actually increased. So the critical indicator is the overall level of traffic in a functional region. In highly urbanised areas, this may necessitate collaboration with neighbouring authorities to avoid the emotional behaviour of one community simply opined with problems, here on traffic problems, to another.

ASSESSING TRANSPORT. POLICIES

With reference to environment monitoring information, the environmental impact of policy is assessed in principle. If the environment is safer, automatically policy would appear to be moving in the right direction, however it is not easy in practice. The state of environment monitoring may not provide a complete data.

The local agenda policy and environmental impact may be distant prospect. However, the state of environment monitoring is essential but not sufficient means of policy appraisal. Two other mechanism can also be tested in this situation, that is the test against good practice and the test of firmness of policy.

GOOD PRACTICE

It can be relatively quick if the good practice tested. There are some useful reviews of the transport field which may be helpful in listing of environment conscious policies from the environmental activists group. The lists examines good practice by mode or transport policy aspect, working from broad transport policy to more specific policies for walking, cycling, public transport, traffic capacity, car restraints, transport interchanges and freight movement.

It is often difficult to define the policy area, and difficult to assess whether it is having elements of beneficial effect. For example, if cycling promotion policy programme is structured by the authority, but where there is no any evidence of implementation either through investment or through development control. Where there is a gap between persuasive language and action, the audit should attempt to pick it up.

Where different environmental criteria forces in different directions, is one another area which is likely to cause further argument. For instance the car parking provision provide for the expected level of car use, to avoid ugliness of the urban area by extra car parking, however, the fact of parking provision itself encourages car use, worse congestion and emission levels and hastens the spiral of public transport decline. Good practice, in this situation, can be very different according to which environmental impact is considered most important.

Conclusively, the good practice test needs to be treated with caution, to ignore conflict between policies and objectives.

POLICY CONSISTENCY TEST

Car sharing

In terms of car sharing and public transport use, policy conflict may arise. Both the option are desirable, as alternatives to low occupancy car use. In practice one tends to undermine the

AN UNIQUE ENVIRONMENTAL BYELAWS

- Bicycles have priority over cars.
- Superinsulation building regulations.
- Domestic dwellings must be fitted with solar collectors.
- A tree must be planted outside every dwelling (to provide shade)
- A tree must be planted next to every parking lot (again for shade).

other. An effective car sharing policy is likely to take passengers away from public transport and thus reduce the visible level of service.

Park and Travel

Park and travel (ride) is an environment friendly policy and it is much understood. At the boundary line of a city, park and travel stations can play a significant role in reducing in-city car use and thus enhancing the city centre environment. But at the same time it is encouraging car use in the outlying areas, deterring people from walking direct to nearby bus services, as a result undermining the viability of local services and contributing to increased energy use. It also inclined to increase the travel some distance to and from work, and so encourage further population dispersals to locations which are entirely car dependent. These examples must be sufficient to indicate that policy assessment is a complex process. It is vital to avoid the rude assumptions – predictable environmentalist responses-that particular policies are automatically good. As a more likely alternative it is the whole context of policy, the overall strategy shape, which will determine whether a particular initiative is worthwhile or not.

The consistency of overall strategy may also be assessed and compatibility of policies pairs to be recorded. Compatibility is examined in relation to over reaching policy goals, in this context 'to reduce transport energy use' and maintaining a good level of accessibility". A particular combination of policies work toward these goals or one policy undermines another can be judged.

CONCLUSION

It is suggested in the above lines that long term sustainability depend on effective integration of land use. Energy and transport plan studies which incorporates extensive survey work with environmental angles should be given due weight. Regular monitoring of transport trends and practices be conducted and it will be useful to move towards strategies Environmental Assessment.

CHECK LISTS

The transport checklist is an attempt to provide a guideline on the environmental audit of the transport.

Reviewing in-house transport practices

Whether the following elements been evaluated:

- choice of vehicles
- quality of maintenance
- use of vehicles for service provisions
- travel by staff for work purposes
- staff journey to work
- whether transport energy use monitored effectively
- whether vehicles selected for purchase or hire, are quiet, fuel sufficient and pollution-limited or free as possible
- Whether maintenance system and contracts set/equipped to fuel efficiency and emission reduction.
- Whether the provisions made for services come into with the necessary number and length of motorised trips.
- Whether work practices encouraged which reduce the need for motorised travel.
- Whether there are incentives to use low energy modes for the journey to work.

Environmental indicators

Whether car use or the overall level of traffic is fall to
city centres
urban areas/rural areas whole geographical region

Whether modal choices for personal travel moving away from energy intensive transport towards walking, cycling and public transport;

for peak central area trips
for intra urban trips
for inter urban trips

Whether emission levels satisfy the standards of Indian and/or WHO standards, in terms of
residential and shopping streets
adjacent to main roads
whether the number of road accidents falling
in absolute terms
in terms of total vehicle miles (by giving focus on levels of severity, different user groups and areas)
Whether levels of traffic noise falling.

- Whether levels of congestion/traffic delay falling.
- Whether bus trip times and reliability improving.
- Whether cycling trip times improving.
- Whether pedestrians delay in crossing main roads reducing.
- Whether violation of environmental capacity in residential areas and shopping streets being stopped.
- Whether the environmental quality trend in terms of traffic is understood.
- Whether a social survey has been undertaken to supplement pressure group views.

TRANSPORT POLICY ASSESSMENT

Capital Transport Investment

- Whether capital programmes moving forward to shift the balance of investment.
- Whether capital programmes (at least 50%) for transport investment linked to complimentary land use policies.

Walking

- Whether extensive pedestrianisation schemes exist in main/central areas.
- Whether a comprehensive pedestrian network is being implemented, designed for convenience, safety and beauty, linked to land use policy.

Cycling

- Whether commuting and recreational cycle routes exist or planned.

- Whether a comprehensive hierarchical route network being implemented/planned for maximum convenience and safety.

Public Transport Priority

- Whether expensive priority on key radials and in the city centre is being implemented.
- Whether effective priority on the whole basic network with all new development designed.

Transport Interchange

- Whether extensive park and travel / bike or ride provision made, or close to commuter settlement.
- Whether comprehensive bike/bus/rail transfer facility provided.

Car Restraints

- Whether there is central parking area for long and short stay.
- Whether general trip-end restrictions on car use exists- in suburbs and exurban area and central area as well.

Traffic Capacity

- Whether restricting traffic capacity and traffic calm conditions in inner urban areas and environmentally sensitive areas follows the effective policies.
- Whether there is a programme of reducing traffic capacity, during the time that giving priority to walking, cycling, and public transport and increasing levels of safety and perceived environmental quality.

Para Transit

- Whether effective measure designed to promote alternatives to the car in low density/dispersed settlements and rural areas, where public transport is not viable- e.g. car sharing, community mini buses, shared taxis.

Freight Transport

- Whether there are implementable policies for freight consolidation and transfer points, linked to rail terminals as well as transport.
- Whether policies designed to ensure that that new freight intensive activities locate on sites where direct access can be gained to motorway/lorry route.

CHAPTER 7

LAND USE PLANNING

This chapter deals with the auditing of land use planning policy. This is the least developed aspect of environmental auditing as it is currently practiced. Usually the planning audit focuses on reasonably separate and identifiable issues that are responsive to some kind of measurement. For example, failure to carry out the protectionary tools for the green belt or the protection of urban green spaces from development. Any or all of these criteria may be important in particular circumstances.

The chapter describes the broader perspective and the reasons for adopting it, then explores some of the very real difficulties which the auditor is likely to have to confront. As a consequence the chapter deals with the land use study, including an urban and rural case study to help give form to the ideas.

Sustainable Development: as new agenda

The relationship between the spatial arrangement of human activities and global environmental quality is charged with conceptual difficulties and preconceptions. The uncertainties of the whole area mean that so far most audits have silently ignored it.

The attitude however, does not avoid answering the responsibility. The interdependence of global ecology, energy use and the form of settlements has long been understood

Energy use is affected by land use planning to a larger extent in a sizable quantity. Built form and layout, density and zoning, the concentration and dispersal of activities cause to very substantial variations in energy use in buildings and transport. High-energy dependence threatens the global environment, in turn. Planning policy guidance revisions have reflected the new priority now, and there is now stress on transport, land use and energy use, and also on shifting the balance from personal car use to public transport, cycling and walking. The government is committed to back the sustainability argument through a system of appeal when appropriate plans incorporate. This, in fact, reinforces the potential significance of the planning policy audit. It relates to test the following two objects:

- Whether the development plan promoted sustainable development.
- Whether ongoing development decisions reflected the environmental priorities in plans.

THE ENVIRONMENTAL APPRAISAL OF DEVELOPMENT PLANS

It involves the following key stages to make the environment auditing clear:

- Monitoring the state of the environment to assess the real impact of policy, and progress towards environmental goals.
- Scoping of the policies in the plan to ensure that they reflect the full environmental/Sustainability agenda and current best practices.
- Evaluation of the policies contained in the plan to assess likely environmental impacts and mutual consistency, leading where appropriate to policy modification and/or measures in mitigation of those impacts.
- As a result EAP requires the planning authorities to be active in the state of the environment surveys, and policy impact assessment. It is practically the main contribution of the planning department or section to environment auditing. It may be as an essential part of the normal process of plan preparation, monitoring and review. The only elements of a complete planning audit which not included in the EAP are the review of develop-ment control and the review of the authorities own development decisions.

PROBLEMS WHILE PURSUING THE AUDIT

It will be worthwhile to discuss some of the problems that restrains progress on incorporating planning in environmental audits before examining the audit elements in details.

The first one is lack of knowledge. More often the planners felt that they have very good understanding of the development issues relating to building and landscape conservation issues, but they are ignorant of the policies which is environmentally sustainable. This tendency to remain unchanged can not be justified or applied longer.

Secondly, the limitation of power to be exercised by the local authority. The decisions of investment companies, multinational companies, government departments and private sector public utility are taken by the central authority/local authority, therefore, are only one of the players in the entire decisive game plan. Similarly, changes in employment of shopping patterns are more concerned with economic and social trends than the local authority policy. Decisions on travel and heating are also made by individual households and enterprises and it is not subject to local authority say.

There may be two answers of these difficulties. One the one hand, while the planners do not make developmental decisions, the behaviour can be controlled individually, it may help in shaping the frame work within which decisions may be taken. If patterns of land use zoning provide a few opportunities for working locally, then it is oblising people to use more energy to get to work. And at the same time hampering social welfare and, possibly, economic growth. Planning policy therefore needs to be geared not to restriction of choice, but increasing choices to the people, allowing them to chose whether to work near home or further a field; whether to walk or use bus or car. It is working on a probable term.

On the other hand, the answer to the plurality of decision makers lies in the nature of the auditing process itself. Ownership of the environmental audit need not be restricted to the local authority. A community audit, linked to an environment concern or local agenda of Rio submit, could draw in some of other powerful agencies and may win their backing.

Problem of making long term commitments may be one another problem.

Thirdly, there may be the problem of making very long term commitments.

One argument against incorporating sustainability into planning policy is the sheer time scale involved in moving from the status quo to a resource efficient settlement pattern.

CHAPTER 8

CONSERVATION AUDIT

The record of designated conservation areas and implementation of conservation policies in the built environment reveals considerable disparities between the approaches of different local authorities. These disparities check to some extent from the discretion given to local authority within the frame work of legislation and government instructions.

Also, it is important for the local authority, elected representatives, local civic societies, residents associations and interested individuals to take stock from time to time and consider whether the conservation work being undertaken in the area. The check points are:

- in terms of an overall conservation strategy, what is required
- whether the conservation work is effective,
- it does not have undesirable outcomes,
- it is aesthetically pleasing
- it has integrity in terms of conservation aims,
- the disparity of policy generation and implementation is probably due to a number of interacting factors such as;
- level of financial sources and well qualified staff
- level of motivation of the officers,
- degree of political will
- lack of pressure from residents and civic societies

However the suggested criteria against which the work of an authority can be assessed during conservation audit:

- record of designated conservation areas and amended record of conservation area boundaries

- notified schemes of preservation and enhancement
- implemented scheme of preservation and enhancement
- the area further being considered as conservation area
- journals of character statements or appraisals for each of the conservation areas
- record of statutory list of buildings of historical and or architectural value
- analysis of the listed buildings in terms of type-style-material and period
- record of town schemes or other conservation programmes and initiatives
- record of delisting or consideration of the de-designation of a conservation area
- record of risk register in terms of building
- record of successful or unsuccessful applications of protectionary measure of the conservation area
- record of special design or zones to be designated as conservation area
- record of special policies to protect skylines, infill, design guides
- record of tree preservation orders
- record of enforcement of building preservation notices, repairs notices etc.
- evidence of commercial sponsorship of conservation schemes awards/competitions for new development in conservation areas
- existence of conservation area development advisory committee
- record of partnership with utilities on siting/coordination and design of utility works in conservation areas
- amount of conservation budget
- use of natural building materials-encouraging opening up of quarry, local bricks or tiles
- quality and quantity of trainees and capable staff

Aesthetic audit

To make aesthetic directives based on the concept of appropriateness to special local identity, in order to maintain a sense of place, it is essential that a local planning authority makes a survey and statement of the character of various areas within its jurisdiction. It must be expected that there is diversity of character across a particular town. The first expectation, thus, in an aesthetic audit that there should be a perceptive appraisal of the existing character of the various areas that comprise the district/area.

The aesthetic experience of an area is not only visual, but it involves other senses; the sense of excitement in discovery of a place as spaces and activities are revealed as one moves through an area such as sounds, smells, tactile experiences. One must be aware of the conditions which contribute to these experiences and the conditions that cause them to disappear in terms of comprehensive redevelopment, change of ownership, gentrification. This is, in fact, managing a sense of a region that gives valuable inventories to sensory experiences.

It is important that the audit is undertaken with representatives from a wider section of society/community, not just the "aesthetically aware-civic society and local authority itself.

The aesthetic audit checklist

- Whether there an emphasis on aesthetic conformity-idea, styles and approaches- or aesthetic diversity
- Whether the aesthetics of various communities acknowledged
- Whether aesthetics regarded as visual aesthetics only or whether attempts are made to enhance other sensory experiences e.g. sounds, smells, levels of vitality
- Whether aesthetic factors significant and well expressed in the purchasing policy
- Whether enhancement schemes planned or in progress in conservation and other areas as well
- Whether any regulatory law prevails to control aesthetic
- Whether there an art policy or programme
- Whether the local festivals/street theatre/events encouraged by the local authority
- Whether there are urban trails prevails to encourage local awareness

CHAPTER 9

POLLUTION CONTROL

In fact, the earth itself is the waste recycling unit. A sizable quantity of pollutants exists in the nature or environment. It passes through the environment, life-cycles and food chains as valuable nutrients to the benefit of organisms and ecosystem without causing damage or harm to the environment, while possessing potential to cause harm.

Discharges of natural compounds from septic tank outfalls into rivers in remote locations can increase the number and diversity of species present whereas a discharge of untreated sewage effluent from a village or town in similar location could eliminate all aerobic life in the river. Unnatural compounds such as chlorofluorocarbons released into the atmosphere will defuse into the stratosphere because of their low solubility in rainfall. In the stratosphere, intense sunlight breaks down the CFC's to liberate free chlorine radicals which are responsible for the destruction of the ozone layer with additional risk of skin cancer at the ground level. Other unnatural compounds such as DDT or PCB's, accumulate through food chains until harmful concentrations are reached, when dispersed into the environment.

Ecosystems also have the ability to adapt to and recover from the effects of pollutants.

Effective environmental management and pollution control, however, requires an holistic understanding of natural environmental processes and the interactions and effects of pollutants to determine whether environmental pollution is taking place alongside the selection of appropriate levels of control. The concept were recognised for the first time in legislation in the environment (preservation) act with a definition of pollution that included the environment as an integrated as a whole:

Pollution of the environment due to the release, to any environmental medium namely the air, water and land, from any process of substances which are capable to harm to man or any other living organisms supported by the environment.

The explanation of harm is critical to the definition of pollution and is taken to mean"

Harm to the health of living organisms or other interference with the ecological systems of which they form part and in the case of man includes causing offence to his senses and harm to property.

Integrated pollution control was introduced for major industrial plants where the environmental option is used to minimise the pollution or to prevent the pollution.

ROLE OF LOCAL AUTHORITIES IN POLLUTION CONTROL

The role of local authority may be divided between rural part of district and the municipal corporation or metropolitan councils. District administration should be made responsible for monitoring levels of air and noise pollution in addition to a range of activities that may be prejudicial to health or give rise to nuisance. The power to control or prevent risks from these activities may be exercised through pollution control regulation may be vested to the competent authority of the district or MC authorities.

Local authority have a valuable role to play in association with the national authorities in relaying information about the local environment or pollution incidents. Local authority can also provide the local base through which the public can gain access to pollution monitoring and control information from other pollution enforcement authority. The policy and activities of the local authority will also have an impact of pollution on the environment.

Local authority, therefore, have a vital role in the control of pollution and any environmental policy of the authority should embrace strategies and initiatives to address local as well as global pollution issues.

Pollution Policy Issues

As per the action programme of the government on the environment, the policies of the local authorities to control pollution in a manner appropriate to sustainable development should contain the global issues and within the philosophy "Think globally and Act locally". Such policies are subject to revise and updated in the light of new findings. Presently, the issues are:

- Air quality,
- Volatile organic compounds,
- Heavy metals,
- Dioxins,
- Acidification,
- Climate change,
- Noise, and,
- Contaminated land.

Air Quality

The government has set air quality standard through a number of directives for SO_2 and particulates, nitrogen dioxide and lead. Ozone is the subject of draft directive. These standards are being restructured and extended under a proposed frame work directive on the assessment and management of air quality with sectoral directives for individual pollutants. These standards are being set to protect and enhance human health, the environment and sensitive ecosystems and also the WHO air quality guidelines for India or Asian countries are likely to be applied. Long term limit values and short term alert limits will be set for each pollutant. In case limit values are exceeded, applicable measures must be taken to reduce concentrations which must be reduced within the limit i.e. 10 to 15 years.

Where alert limits are exceeded, the public must be informed.

Standards for sulpherdioxide, nitrogen oxides, black smoke, suspended particulates, lead and ozone will be established on certain previous level. Standards for carbon monoxide, acid deposition, benzene, aromatic polycyclic hydrocarbons, arsenic, fluoride and nickel will also be established on a certain previous level.

Local authority may have a powers under the Environment (Preservation) Act to investigate circumstances that may be prejudicial to health or a nuisance. Besides that, powers are available in the clean air act to undertake or fund research relevant to the subject of air pollution. Monitoring should also be taken to ascertain compliance with for quality standards and guidelines, to identify the particular sources of air pollution and also to assist in the development of control strategy through a range of measures such as location, process controls and emission controls.

VOLATILE ORGANIC COMPOUNDS

VOC's symbolise a wide range of organic compounds from many sources such as combustion, transport, incineration, and use of solvents in industrial process. Certain VOC's are toxic and carcinogenic while others are involved in the formation of ozone through complex reactions between oxides of nitrogen and sunlight.

The govt. may have targets for reduction of manmade emissions of VOC's. This will be achieved by drastic controls on emissions from industry through the proposed directive on VOC's and by additional controls on transport. It is important that the emphasis should be given to reduce and eliminate the compounds which are classified as carcinogens, mutagens and toxic to reproduction as well as those that are chlorinated. Local authorities should be monitoring ambient air quality to identify where air control guidelines or standards are being exceeded.

HEAVY METALS

The government may have set a target of at least 70% reduction from all courses of cadmium, mercury and lead emissions. This will be achieved by control of emissions at source by substituting lead free petrol or by more strict controls on emissions.

Local authorities may investigate all sources of heavy metals in their area and establish the applicable environmental options to control releases. Where metals are releases, monitoring should be undertaken to ensure air and water quality standards.

Dioxins

Dioxins are subjected to strict controls and the government may have set targets for reduction of dioxin emissions from identified sources.

These compounds could be produced during the combustion of chlorinated compounds in municipal, clinical, hazardous waste incineration processes in addition to cremation. The government may set a target for reduction of dioxin emissions from identified sources.

Local authorities within their jurisdiction should cooperate to establish a policy that adopts the possible environmental options on the disposal of such wastes. This may involve a more efficient way of incineration facilities and pretreatment of wastes to ensure that incinerators are provided best environmental option and are operated under optimum combustion conditions so that the destruction of these compounds can be ensured. Alternatives to such incineration should also be considered, e.g. sterlisation of certain clinical wastes with disposal by landfill.

Acidification

Combustion of fossil fuels and agricultural practices are changing pH of the soil causing mobilisation of toxic metals into surface and ground water. Critical loads should be calculated for the deposition of acid, either sulphates or nitrates or alkalies to identify soils that are at risk.

The government may set the aim for not exceeding the critical loads and to achieve this aim, emissions of nitrogen oxides and SO_2 should be stabilised at certain previous level.

Local authorities should assess their areas to ascertain whether critical loads are being exceeded or whether emissions from their areas are causing critical loads to be exceeded in other areas. Policies for controlling such emissions should be developed, e.g. transport, the nature of industrial development, energy efficiency and the use of appropriate fuels.

Climate change

Releases of greenhouse gases could put problems through global warming which may effect meteorological patterns, climate, flooding and agriculture. Targets may have been set within the government to stabilise certain levels of CO_2 emission and to phase out CFC's and other ozone depleting substances and to gather information of methane and nitrous oxide releases.

In fact, actual emissions of CO_2 may be little known within the boundary of local authority from industry, transport, domestic and commercial sources. Emissions inventories should be established at the side of other atmospheric pollutants and control strategy devises.

Noise

The local authority may administer and control of noise effectively which will involve a dialogue with the environmental health departments to separate noise sensitive developments from noise generating activities. Besides this, sufficient sources must be available to establish current background noise levels and existing noise sources by a range of educational, legislative and technological approaches. The local authority should act to control the noise pollution and also may take necessary steps to follow the minimum standard set for noise.

Contaminated Land

Government policy, while dealing with contaminated land, may be as follows:

- Prevention or minimization of further pollution with the polluter paying for any damage or necessary controls.
- Dealing with existing contamination which poses actual or suspected risks to health or the environment.
- Improving sites to a level, suitable for the proposed land use.
- Encouraging the development of land subject to potential or actual contamination provided appropriate investigations and remediation work are carried out.

District administration have an important role to play in controlling the redevelopment of contaminated land and in preventing new developments that may contaminate land. Proposals for the possible reclamation and use of contaminated land can be included in development plans and government grants may be available for the cost of investigations and remediation. Controls on the redevelopment of contaminated land depend on identification, and the remediation appropriate to the land use, or restrictions on land use dependent upon the level of contamination.

Local authorities should continue to gather information on the historical use of land as an indication of, whether contamination may have taken place. Where such evidences exist, risk appraisal to health and of damage to the environment should be carried out with appropriate site surveys. Priorities should be made for safety of contaminated sites and remediating sites to a level appropriate to current use or proposed use with the responsibility on polluters, owners, or developers to undertake. Where contaminated land is in public ownership, funding should be sought to enable remediation and beneficial to development of the land to take place in preference to new development on·green field sites.

WATER

Local authority have a definite role to play in controlling drinking and bathing water quality, in preventing pollution of surface waters and ground water, the restoration of natural ground water and surface waters to an ecologically sound conditions, and in ensuring a balance between the water demand and water supply on the basis of rational use and management of water resources. Local authorities are also the user of water and through their activities, it will contribute to the overall load on sewage treatment works as well as surface run-off into water courses. The use of biocides is also likely to cause a deterioration of water quality.

Planning control over the use of land for the purposes of housing, roads, industry, agriculture or tourism will affect the demand for water and the nature and quality of discharges into aquatic environment. The establishment and achievement of statutory water quality objectives for improvements in ground and river water quality will require cooperation between the various pollution enforcement agencies with pollution control strategy or possible environmental option.

Overview

So far pollution policy is concerned, priority should be set for relevant issues with objectives, targets and timescales depending upon local conditions and circumstances. Proper resources and personnel must be provided to implement the pollution control policies. The responsibility and relationship of the key personnel must be defined and documented. There should also be effective communication with staff, councillors and the public. Training programme for relevant personnel should also be arranged.

POLLUTION CONTROL AUDITS

State of the Environment Audit: to assess information on the natural environment to the side of pollutant emissions and the achievement of environmental quality standards

Pollution management studies: to review and assess the effectiveness of policies of the authority in relation to initiatives dealing with pollution and the enforcement and effectiveness of legislation in order to control pollution.

STATE OF THE ENVIRONMENT AUDIT

An understanding of natural environment and its interactions and effects of pollutants on human beings and other living organisms supported by the environment is required for effective pollution control and environment management. If formation on the nature and quality of pollutants emitted is also necessary to assess likely environment levels and trends in emissions in details. Modelling of emissions can be carried out in association with meteorological, hydrological and geological data to predict the dispersion or propagation of pollutants in the environment. As the total outcome of environment levels will also depend on removal mechanisms and other interactions such as solubility, adsorption, biodegradation, and chemical and photochemical reactions to be incorporated into the models. Local environmental monitoring should be used to validate the models as well as establishing compliance with environmental quality standards and guidelines. Environmental risk assessment should also be incorporated into the setting of environmental targets and objectives in the management of local environment quality. The application of this information includes:

BEST PRACTICABLE ENVIRONMENTAL OPTION

In establishing the best practicable environmental option, the disposal route (air, water or land) must be selected only after a detailed analysis of options has been made including steps necessary to reduce waste to the minimum practicable. BPEO should be considered at the initial stages of the development of projects. It need not be complex but should be comprehensive, dealing with the whole environment, anticipating effects of dischargs on the environment and considering the control of pollution by a range of technologies recycling and waste disposal. The preferred option should be examined in detail and the cost of pollution control balanced against the protection of the environment. An audit trail should be used to record the decision process and decisions should be open to scrutiny. Finally, the selection of BPEO should be kept under review as circumstances, change.

Source : 5th Report of the Royal Commission on Environmental Pollution, 1975.

Identifying the source of industrial emissions causing a nuisance in the neighbourhood.

Determining proper levels of control in existing pollution problems such as concentrations on oxides of nitrogen from motor vehicle response to accidental situations by the release of hazardous materials into the environment at potentially harmful concentrations.

Appraisal of proposals for new development in regards to the effects on the environment.

Investigating epidemics with a view to establish links with pollution sources.

Local authorities are supposed to provide annual reports on air, water quality, noise and contaminated land.

On the national level, the annual report of the MoE/F provides an overview on the environmental quality and natural resources

INFORMATION NECESSARY TO PROVIDE AN EFFECTIVE POLLUTION CONTROL AND ENVIRONMENTAL MANAGEMENT SERVICE

Weather data

To be used for air pollution dispersion models, energy demand and water quality models :

- Wind speed and direction
- Rainfall
- Temperature
- Humidity
- Atmospheric pressure
- Stability class
- Solar radiation including UV-B

Air Pollution

Sources

To prepare emission inventories and predict ground level concentrations through dispersion modelling :

- Authorization
- Emission limits
- Chimney heights
- Discharge parameters :
  - temperature
  - volume
  - efflux velocity
  - concentration of pollutants
  - mass emission

Environmental levels

To ascertain compliance with air quality standards and guidelines, to investigate and establish sources of air pollution and to gather data for epidemiological studies :

- Total particulates
- Respirable particulates (PM 10)
- Oxides of sulphur
- Oxides of nitrogen
- Volatile organic compounds (VOCs)
- Heavy metals
- Radon and other gases
- Odours

of the country, which could be replicated at the local level. The Indian government has also framed pollution control policies within the strategy of sustainable development. These documents provide the basis for environmental policies on which the targets can be set. A similar approach should be adopted at the local level. The local authorities should establish their role as custodians of the local environment.

Necessary Information needed to provide an effective pollution control and environment management service.

WEATHER DATA

Weather data can be used for air pollution dispersion models, energy demand and water quality models:

- Wind speed and direction
- Temperature
- Humidity
- Atmospheric pressure
- Rainfall
- Stability group
- Solar radiation (including UV-B)

AIR POLLUTION

Sources can be used in preparing emission inventories and predict ground level concentrations through dispersal modelling:

- Emission limits
- Chimney heights
- Discharge parameters in terms of
- temperature
- volume
- concentration of pollutants
- flowing velocity
- authorisation

Environmental levels is required to ascertain compliance with air quality standards and guidelines, to investigate and establish sources of air pollution and to gather data for epidemiological studies:

- Total particulates
- respirable particulates
- oxides of sulphur

- oxides of nitrogen
- Volatile organic compounds
- heavy metals
- radon and other gases
- odours

Dispersal models can be used to predict the likely consequences/effects of proposed developments or accidents on environmental quality:

- Source of the point-line and area
- Ambient air quality

NOISE POLLUTION

Sources can be used to monitor compliance with standards and establish measures to remedy any nuisances or excessive levels:

Industrial
Commercial
Domestic
Motor vehicles
Railways
Airports
Leisure hour activities

Environmental levels is required to establish background levels and prevent any deterioration in environmental quality in terms of noise:

Models can be used to predict the likely effects of proposed developments on environmental quality by noise propagation and reducing the noise by law:

Construction
Industry
Roads
Railway yards
Airports

WATER POLLUTION

To provide information on resources, water quality and sources of pollutants:

Natural river water quality
Natural ground water quality

Sources of pollution
agricultural
industrial
domestic
Discharges
location
discharge parameters
temperature
volume
concentration of pollutants
River water quality
Ground water quality
Bathing water quality
Drinking water quality
Models
Dispersion in ground water, rivers and estuaries

LAND POLLUTION

To provide information on ground water resources, quality and flow, contaminated land and suitability of land use:

Natural geology
characteristics
permeability
ground water flow
Contaminated land
location
type of industry/activities
date/duration of activities/contamination
significance of contamination whether low/moderate/or, high
nature of pollutants
asbestos
metals ad metalloids
inorganic anions
organic materials/compound
properties of pollutants
pH/solubility/redox potential
chemical reactions-synergism/antagonism
persistence
evaporation / dust potential

adsorption / absorption
degradation / production of methane and other gases
likely effects
air quality
fire / explosion risk
water quality
damage of materials
bio-accumulation
toxicity to humans and ecosystems

POLLUTION MANAGEMENT AUDIT

The effectiveness of policies of the authorities in regard to initiatives taken for pollution control and the enforcement and effectiveness of legislation controlling pollution should be reviewed and assessed by Pollution management audit. The quality control standard provides a specification for environmental management system to demonstrate compliance with a stated environmental policy.

The elements of the environmental management system can be applied to the enforcement activities and initiatives of local authorities related to pollution control, providing an indication of policies, procedures, activities and actions to be audited.

The objectives, scope and execution of the audit should be clearly defined. Areas likely to be covered by the audit includes:

Organisation

organisation structures and responsibilities to implement pollution control policies and initiatives including relationships between different departments, committees or agencies

Knowledge

understanding and knowledge of legislative controls, standards, guidelines and targets
understanding and knowledge of the environment, ecosystems and the interactions between individual sectors.
Administrative and operational procedures
Management procedures and manuals, operational controls, production of records and reports, quality assurance

Environmental performance

compliance with legislation, emission and environmental quality standards/targets

effectiveness of surveys, assessment and evaluation of monitored data

use of environmental models in pollution control and environmental quality management

enforcement action, response to complaints and incidents including emergency situations

Human, physical and financial resources

sufficient staff, training and qualifications

monitoring and control equipment, access to information and databases, facilities provided at worksite and office.

Communications

sufficient provision for current and anticipated legislative requirements and other environment friendly initiatives.

To identify difficulties or deficiencies in implementing the pollution management system or in collecting, assessing and evaluating environmental information is the main function of the environment audit. The audit may note the problems in achieving the environmental standards and targets, and may suggest remedial measures.

The policy on environment will have to be reviewed to establish its continuing effectiveness. As a part of the review process the following matters to be addressed:

Any recommendations which have been included in the audit reports, and also the way of implementation of the recommended parts.

The suitability of the environmental policy in continuity, and revisions in the light of followings:

- growing environmental concern in specific areas
- developing understanding of environmental issues
- potential regulatory developments
- concerns among interested parties
- market pressures
- changing activities of the authorities and,

- changes in sensitivity of the environment
- the suitability of environmental objectives and targets, and any revisions to the environmental programmes

CHECKLIST

The checklist shapes the conclusion and the summary of this chapter.

- Whether the precautionary principle applied?
- Whether environmental effects are being taken in to account in decision making at the early stage?
- Whether the exploitation of nature or natural resources being avoided which causes damage to ecological balance/
- Whether scientific knowledge has been improved to make possible the appropriate action to be taken?
- Whether polluter pays?
- Whether the general people has concern over protection of environment?
- Whether relevant data on environment is readily available?
- Whether the appropriate control level / system is being applied?
- Whether sustainable development approach is being strengthened ?

Pollution checklist

Air quality

- Whether monitoring is being conducted to ensure compliance with national quality standards for SO_2 & particulates, NO_2 and lead ?
- Whether WHO guidelines are being applied for organic and inorganic compounds, as well as considering implications for sensitive ecosystems?
- Whether sources of cadmium, mercury and lead and other such heavy metals have been investigated and monitored ?
- Whether environmental options have been considered, where metals are released?
- Whether there are policies to monitor and control emissions of toxic metals caused by he combustion of fossil fuels?

- and agricultural practices which caused acidification?
- Whether the required information collected, including weather data, to assess the nature and quantity of pollution emitted?
- Whether potential trends for epidemiological studies?

Water Pollution

- Whether local authority role is given weight in water consumption and management through its own activities, and as also a role model for reducing consumption?
- Whether sufficient information provided to the public on the quality and control of water sources?

LAND POLLUTION

- whether all proposals for new development are appraised in relation to possible effects on the environment, which may include ground water resources, natural geological factors and existing soil contamination.

NOISE POLLUTION

- Whether all the potential sources of noise pollution have been established and monitored to confirm with the national targets.
- Whether sufficient liaison carried out between the planning and the environmental health deptt. To ensure that noise sensitive developments are separated from noise generating

Chapter 10

WASTE AND RECYCLING

Auditors working on waste and recycling may face difficulties, which arises due to the understanding that the recycling issues are of short-term issues. Because, there are no well-spelt definitions of waste categories, inter-authority comparisons can be meaningless. As producers of waste, most people regard themselves as an expert, and nonsense talk about waste materials continues. However, the auditors can be supposed to encourage local authorities or councillors to follow the traditions established by many of the practising pioneers in terms of the uses of waste disposal techniques or may be followed on the dictations of such apex body under the central government. It can be expected to take the wider view in relation to weighing national needs against local considerations. And, the auditors should be aware that it is necessary to pursue their enquiries all the way if they really wish to identify or to get the root of specific disposal policies. The need for guidance on the relationship between planning and pollution control in the light of new measures in the environmental protection act.

Presently, the proportion of country's annual waste which is collected or disposed of by the local authorities is very small and it has no any recycling system or management. The authority is needed to change the priority over waste disposal, waste reuse, waste recycle and its commitment towards sustainability, as a much more fundamental point.

Local authority still have an important role to play in the management of solid wastes and cleansing of streets and public places, as they are responsible for ensuring that million tonnes, out of which more or less 85% of the total generated waste are being collected from dwelling in the country per year. The central government is also keen to initiate its role in imposing legal standards for the performance of the service, and becoming increasingly involved in setting standards.

There is an evidence of damage that results from litter and roadside dumping. Auditors can ascertain whether the obligations set out in the civic amenity legislation to provide facilities for recycling opportunities, its reuses or disposal.

REFUSE COLLECTION

The way and routine the refuses put in the dustbins by the house holders, collected by dustmen and manually thrown into dustcards is time honoured job. The domestic refuse collection routine often requires dustbins and sacks to stand in the streets ahead of collection vehicles, and the consequential activities of scavengers have often resulted in the scattering of waste around pavements.

The use of plastic rubbish sacks/bags

It has no doubt that the plastic bags are useful in handling refuse and breaking the life-cycle of the housefly. At the same time, the plastic bags are not inherent environment friendly as they consume non-renewable, and, are not usable or readily recyclable and they also have disadvantages within a landfill once tipped. Biodegradable and photodegradable bags which breakdown more readily in the environment are available.

A good alternative to the plastic bags is the wheeled bin, which can be mechanically loaded into the collection vehicles. It does offer solid environmental disadvantages, such as the diminished use of non-renewable source and the recycling of plastic from scrapped bins. While benefits also arise from reductions in street litter from burst bags, and improvement from the health and safety of the refuse collectors. The generous capacity of the "wheel bins" chosen by local authority has led to criticism that they encourage profligacy in domestic waste production.

Auditors should ask whether the council/LA has a policy on the use of electric sink waste grinder and/or pipeline system. If the electric sink waste grinder used in large numbers, these latter devices not only consume extra energy in the grinding process, but also contribute to the loading on domestic sewage disposal works to the detriment of local watercourse quantity.

COLLECTION OF INDUSTRIAL AND COMMERCIAL WASTE

Most of the industrial and commercial wastes are collected by private enterprise with local authorities collecting only nominal amounts. Due to the lack of direct interest from municipal authorities, facilities for disposal of such wastes have been limited and a tradition of unlicensed dumping has arisen, notably among the small companies. Auditors should enquire into the extent to which local authorities are enforcing the duty of care provisions imposed under the EPA on waste producers. These duties include the applicable measures:

- to prevent the escape of the waste,
- to prevent any unlawful disposal of the waste by any other person.
- To secure (on the transfer of the waste)
- that the transfer in only to an authorised person
- that there is transferred such a written description of the waste as, will enable other persons to dispose of the waste lawfully and to comply with the duty to prevent the escape of waste.

Many environmental problems caused by waste arise from mere carelessness or casual storage or thoughtless disposal of waste. In fact, awareness or education is must to ensure that the packaging is good enough, not merely for the next link in the disposal chain, but for the entire process to delivery at the point of final disposal. Such obligations already exist in a similar form in the regulations governing the carriage of dangerous substances by road.

Disposal of Waste

Landfill: Auditors should find and ascertain the extent to which resources give over to the monitoring of conditions on site for the suppression of nuisance from smell, dust, airborne paper and various pests, and also the aftereffects of landfill gas and leachate (leachate is the offensive liquid which escapes from landfill sites due to rainfall washing out organic pollutants and metals from the wastes) generated on the sites.

The auditors should also seek information whether the authority has undertaken feasibility studies on the utilisation of methane generated during the landfill decomposition process as an alternative energy source. Such exercises can be financially viable on many sites.

Incineration

Incineration as a concept has undergone a significant change from the change when cities or towns possessed several smoky and smelly manually fired "destructors-which were thought to be necessary to sterilize domestic waste. Now a day incinerator is likely to be a sophisticated continue fireplace, which may boast beat recuperation, even electricity generation, and incinerator manufacturers extensively promote this expensive hardware. Certain technical difficulties, however, have force against easy access to the considerable heat potential of burning domestic refuses. These difficulties centrally points on the fact that its diverse nature makes it a far from actual fuel, with corrosion and slagging of the heat exchange surfaces also imposing significant cost penalties.

Landfill gas utilization uses

Gas from a landfilled claypit used to fire housebricks and power 1 MW generators via spark ignition engines.

A scheme in which gas originally powered a boiler which was raising steam for energy hungry cardboard manufacture.

Nursery developed close to a small landfill to demonstrate the feasibility of small scale operation using gas to heat greenhouses in Yorkshire.

3.7 MW of electricity exported to the local grid in the vicinity of TP stations.

Gas (collected by a grant-aided pipeline system) generates 3.8 MW of electricity via Centrex gas turbines.

Other options

Auditors need to be careful that although many ingenious (and sometimes outrageous) alternatives to the landfilling and incineration of domestic waste have been tried, almost every attempt to scale up such processes from pilot plant to workable proportions seems destined to throw up a host of technical and economic difficulties. The production of refuse-derived fuel (with a nuisance free, transportable product) and the destruction of waste by pyrolysis (which yields a rich hydrocarbon feedstock) both provide recent examples of promising ideas, which have not yet achieved commercial success. Hence no panacea is in immediate prospect which short-term market considerations continue to predominate the decision making process.

HAZARDOUS WASTE MANAGEMENT

The not in my backyard syndrome will usually mean that any disposal site which treats 'special' wastes tends to arouse fears in the minds of local residents, these (sometimes irrational) fears often being expressed via virulent public protest campaigns. Hence applications for planning permission are often refused or (like many other waste disposal site license applications) have to be settled on appeal by the Environment authorities.

Although the appalling record of some hazardous waste operations has quite properly been a historical cause for concern, vastly improved standards of regulation and handling do now allow a more rational analysis of the dangers. This should facilitate the making of decisions based on an objective assessment of the hazards from such an undertaking alongside the risks from other companies, which, for example, may be handling virgin hazardous chemicals. The current economic situation is forcing communities to become more objective bout their fears; nowadays the fact that a regional disposal facility might help to attract much needed industrial development to an area of high unemployment may induce a groundswell of support for the proposal.

RECYCLING OF WASTE MATERIALS

In fact, many of the earth's resources (notably minerals and metals) are demonstrably finite, should mean that it is easy to convince people that it is imperative and prioritise their own consumption patterns to take account of this fact. Unfortunately a whole panoply of forces constrain this noble goal (not the least of which is the short timescale on which most politicians operate). Hence, certainly within local government, sustainable economics (and even waste minimization) excite much less attention than recycling of wastes.

Surveys of public opinion on green issue show that matters related to waste (and especially recycling) are now perceived to be of significant importance. The concept of credits claimable by the recycling agents (introduced in the EPA) should help to focus attention on these topics, and few authorities are nowadays ignoring the opportunities which exist for obtaining media coverage for their latest recycling activity, however this may be inconsequential. The lack of real interest by Las in sustainability has been a common focus of

discontent amongst green groups (but auditors should note that these are beginning to alter this)

Local authority recycling targets

Recycling is mentioned many times in representations made by the public to any auditors and although certain Las are currently seeking a high recycling profile, councils are frequently still criticized for failing to take an adequate lead in this respect.

Auditors should note that at least one district council or LAs may publish its plan on the grounds that the resources were not available to implement it. Some of the Rural District Councils may encourage recycling activities and rag pickers faces greater logistical problems in the countryside than in a densely populated in promoting recycling in rural communities of course, but these difficulties are not unsurmountable.

Education regarding recycling

Media campaigns can change people's attitudes. However it is significant that market mechanisms proved incapable of supporting the same service in peacetime. LAs are unlikely to find space for the development of any significant recycling schemes which show a sensible return on investment, as all materials which can be profitably recycled (such as ferrous and precious metals), will already be being collected by the private sector.

Although it is unrealistic to assume that it is feasible to return to the feeding of domestic kitchen scraps to livestock (e.g. the domestic fowl), the cessation of this practice has obviously increased the waste of these valuable resources, and also contributed to the worsening of the nuisance inherent in landfilling or incinerating waste (which nowadays has a higher proportion of putrescible domestic discards). Thereafter, auditors should encourage authorities to minimize the output of decomposable waste from all houses by encouraging those with gardens to compost such wastes along with garden trimmings, and assisting others to use small domestic wormeries (or slug bins) into which kitchen food scraps can be placed for recycling.

Examples of recycling

Certain LAs may have a long commitment to recycling and the last few days have witnessed a rapid expansion of interest in the urban/metros and that should be extended to Las on the part of Las in the topic of recycling.

LAs may respond in some fashion to the statutory imperative for recycling. It is the task of the auditors to ascertain the extent to which this commitment is really productive.

CHECKLIST

Attitudes towards waste disposal

- Whether waste disposal has been taken into account in development control by the planning authority?
- Whether environmental assessments have been required for all major waste disposal applications?
- Whether environmental auditors are pressurizing politicians and officers to think more radically than simple waste management and recycling schemes?

Refuse Collection

- Whether the LA's obligations are under Civic Amenities Legislation fully discharged?

- Whether the council debated has the environmental impact of any plastic bin linters which they purchase and what would be the cost benefit aspects of switching to the wheeled bin?
- Whether they have policies on sink waste grinders?
- Whether they have examined pipeline removal from their urban centre? If it is in practice?

Garden Waste

- Whether advice is given to householders on composing? If yes, to what extent?
- Whether the LA has a policy of encouraging composting?
- Is finance being sought to follow concepts of the 'chip don't tip'?
- Whether the authorities, by addition of wood chips to compost, have been able to completely eradicate their own use of peat?

Industrial and Commercial Waste

- Whether the prompt care in respect of waste being enforced rigorously?
- Whether waste production volume and composition criteria are used by planning officers When vetting proposals for new industrial developments?
- Whether authorities assist local skip handling firms in the establishment of transfer facilities (thereby permitting the more efficient transportation of wastes to environmentally acceptable locations)?
- Whether local Councils engaging in 'Waste Awareness' programmes to draw the attention of the public and industry to the need to minimize waste production?

Incineration

- Whether the disposal options have been objectively compared using cost benefit analysis?
- Whether methane capture being practised on landfill sites?
- Whether a liaison committee exists to facilitate the LSG/ Councils getting together at the operational level to review the availability of disposal facilities in their area and the implications for recycling practice?

Hazardous Waste

- Whether there are advantages in having a local special waste facility which avoids the necessity of transporting wastes over unacceptably long distances?
- Whether the question of a regional special waste disposal facility has been fully explored?
- Whether the media are fully briefed and using objective criteria as to the actual extent of the hazard from special waste disposal?

Attitudes to recycling

- Whether the Council/LSG being put under pressure to be proactive regarding sustainability?
- Whether it is using all possible opportunity for achieving improvements on the personal and institutional front?
- Whether some more efforts can done to increase the extent of recycling by schools, uniformed Organisations, public/ charity organisations/NGO's etc.
- Whether the LA/LSG encourages (or provide) wormeries?

General Considerations

- Whether all of the key issues have identified above been properly examined?
- Whether local authorities cope with the global dimension inherent in questions related to sustainability and waste management?
- Whether the political dimension is being fully explored, viz. are local councillors fully alive to the national dimensions of local decisions on waste disposal and recycling?
- Whether recycling frames as salvation or just a smokescreen?
- Whether local authorities are able to persuade local industrialists to take their own wastes seriously?
- Whether the Council/LSG is fully complying with the various performance standards specified in legislation and ministerial edicts (these include litter removal and recycling targets also)?
- Whether the Council undertake the statutorily prescribed 'duty of care' regarding its own wastes?

Chapter 11

ECO-CONSUMERISM—THE PURCHASE AUDIT

INTRODUCTION

The state of the environment, whether on a global or local level, is now such that everyone must participate if we are to halt and reverse its decline. Consumers have a vital role to play in that process.

Today there is more environment information than ever before about the products we buy. There are environmental claims attached to almost all-modern goods: less packaging, low energy, biodegradable, recycled, CFC frees and so on. In spite of this amount of information however (or may be because of it), purchasers often find it difficult to make choices between the environmental impact of one purchase against another. Information is often presented in so many different ways that comparisons between the environmental impact of one purchase against another. Information is often presented in so many different ways that comparisons between purchases are at best difficult and at worst, impossible. Identifying goods with the minimum environmental impact is therefore not easy.

This is hardcore fact that, no manufactured goods are without an environmental impact. Consequently, the purchase of products and services is an important part of an environmental audit. Purchasing may be of the key issues addressed by many local authority audits. Local authorities purchase large quantities and many types of products and services in the course of their activities and therefore have the opportunity to reduce their environmental impact across many different types of goods and services. This includes the purchase of building materials, food, energy, vehicles, road materials, school and office supplies, steel furniture, cleaning materials and horticultural products. This chapter is concerned with the process of evaluating and comparing the environmental impact of goods and services purchased by local authorities, and identifying ways of reducing these environmental impacts. It may be divided into four parts, dealing in turn with the methodology of the purchase audit, the principles for developing environmental purchasing policies, the effect of green consumerism on purchasing and the environment, and the terms of reference for a purchase audit.

PURCHASE AUDIT METHODOLOGY

The purchase audit provides a way of assessing and implementing environmental checks and controls on goods and services. The process can also assist the development of environmental purchasing policies and raise awareness about the environmental impact of purchases in the organisation.

The purchase audit is a key part of the internal audit and involves the assessment of the environmental impact of all the organisational policies and practices that relate to the purchase of goods and services. The purchasing audit is aimed that

(1) to assess the environmental costs and benefits of each discrete policy and practice (but not to place monetary values against these);

(2) To provide a comprehensive source of quantified environmental data and analysis to guide the development of revision of policy and practice.

All purchases should be itemised and the significant environmental impact of which should be identified. Significance of environmental impacts may be measured in terms of environmental importance or volume. Thus, a purchase has a significant environmental impact, the environmental effect is important even if the volume of purchasing is small (such as products containing CFCs), or if the volume of purchasing is large even if the environmental effect is less important (such as paper products). Purchases of goods made through contracts for services should also be identified.

For each environmentally significant purchase, the following measurements should be made in the internal audit:

- the quantity/volume of products purchased (tonnes, reams, litres, etc.):
- the amount of money spent on these products; and
- the proportion of less environmentally damaging (environment friendly) products purchased

With periodic assessment, these measurements may allow the examination of purchasing trends over time, and also can form the basis of targets for the absolute reduction in consumption of damaging products and targets for the proportional increase in consumption of less environmentally damaging products.

Purchasing responsibilities should also be established as part of the purchasing audit. This involves identifying for each environmentally significant purchase the person(s) responsible for the following tasks:

- deciding on the type, quantity and source of products purchased;
- researching, earmarking or identifying less environmentally damaging products; and
- measuring, monitoring, enforcing and reporting on any purchasing policies and targets.

In local authorities, responsibility for purchasing may split between operational units and departments. The following issues affecting the management and control of purchases in an organisation should also be examined in the purchase audit:

- the level of centralisation over purchasing decisions about the type, quantity and source of products – the more decentralised the decisions, the more difficult it is to control purchasing according to environmental (and other) critertia;
- the number of people with the authority to make purchasing decisions – the greater the number of people involved, the

more difficult it is to monitor and enforce purchasing policies; and

- the role of any central supplies, department (or equivalent) and how proactive it is, in earmarking or researching and promoting the purchase of comparatively less environmentally damaging products.

These issues influence the way in which the organisation is able to deal with environmental issues related to purchasing. They are also important issues when developing environmental purchasing policies.

PRINCIPLES FOR DEVELOPING ENVIRONMENTAL PURCHASING POLICIES

Introduction

An environment friendly purchasing policy addresses two key issues: minimizing the depletion of non-renewable resources, and avoiding the creation of pollution. Effective environment friendly purchasing policies should be based on an effective environmental strategy and sound environmental objectives, and should take into account the purchasing process of the local authority. Policies should be applicable to all purchasing activities in the organisation and should draw on the responsibility of everyone involved with the organisation, and emphasize the contribution made to the improvement of the environment on both a local and global scale. The main role and the objectives of an environmental; purchasing policy are summarized in boxes.

Purchasing Process

There are a number of key questions about the purchasing process to be asked before developing an environment friendly purchasing policy.

- Whether there are already any environmental policies relevant to the purchasing process?
- Whether the personnel are aware of existing environment friendly purchasing policies?
- Whether the responsibility for purchasing and product specification has taken by someone.

- Whether the organisational purchasing structure was made?
- Whether there are information sources held about environmental purchasing (internal or external to the organisation)?
- Whether experienced officers (such as trading standards, environmental health, legal or health and Safety personnel) are involved in the purchasing process?

Policy Content

Because the local authorities should have environmental purchasing policies, a abroad look at some of these may provide a guide for policy development. Regional or organisational variations mean that policies should not simply be copied, however – the language and form of the policies should reflect the particular authority. The policy document may contain eight product groups (chemicals; construction materials; energy; peat; plastics; stationery; vehicles and fuels; and woods and metals) an inventory of the products purchased, the environmental issues associated with them, the relevant LSG council objectives, policy recommendations, a good practice guide and an action plan for implementing and monitoring the policies (comprising recommended action, lead officer(s), implementation timetable and performance indicators).

The main tasks of an environmental purchasing policy

- To discourage unnecessary purchasing and consumption of products and services
- To discourage the use of goods and services that heavily deplete non-renewable resources (either in their production, distribution, use or disposal)
- To encourage the use (and reuse) of goods and services that minimize the depletion of non-
- Renewable resources
- To discourage the use of goods and services that cause excessive pollution (either in their Production, distribution, use or disposal)
- To encourage the use (and reuse) of goods and services that minimize pollution.

Objectives of the environmental purchasing policy

- Reduction in consumption wherever possible
- Use of products more intensively
- Purchase of products manufactured from recycled or renewable materials
- Replacement of more environmentally damaging purchases with least environmentally damaging purchases
- Purchase of products without unnecessary packaging
- Purchase of long-lasting, durable, reusable, refillable, repairable or recyclable products in preference to short-life, insubstantial, disposable alternatives
- Active search for the least environmentally damaging purchases
- Promotion of the use of the least damaging purchases
- Reduction in waste arisings
- Recycling and reuse of waste arisings wherever practicable

Compulsory competitive tendering

A number of local authority activities are now subject to compulsory competitive tendering (CCT). While local authorities may have restricted powers to impose environmental conditions on contractors, they nevertheless have a general duty to set standards for products under consumer legislation. They may also have a general duty to take reasonable care and apply expertise to ensure that goods are fit for their purpose and that services conform to the customer's requirements. Department of the Environment Circular may entrust or advice local authorities that to follow the criteria in this area and also state that tender documents must avoid naming particular products and avoid descriptions or specifications which point to or favour a particular suppler. The use of technical specifications which mention articles of a specific make or source, or produced by a particular process, and which therefore favour or eliminate certain alternatives is also prohibited. It is, however, possible to specify compliance with Indian or SAARC. The Local Government Act may prohibits the inclusion of 'non-commercial matters' in contracts. Many environmental criteria, however, may legitimately be described as commercial local authorities may have establish environmental criteria , however may legitimately be described as commercial. Local authorities may have establish environmental criteria in their purchasing policies.

Life cycle analysis

Environment friendly purchasing should be based on the selection of products with the least environmental impact during the entire life cycle of the product. This approach, known as life cycle analysis (LCA) may be used to assess the environmental impacts of products for the Indian countries ecolabel.

Management Issues

Developing an effective environmental friendly purchasing policy may involve the creation of a small interdepartmental team whose responsibility as to may be defined identify problem areas within the

Definition of life cycle analysis

Life Cycle Analysis (LCA) involves carrying out an inventory of the main impacts associated with the manufacture, use and disposal of a product, from the mining of the raw materials and energy used in its production and distribution, through to its use, possible re-use or recycling, and eventual disposal. A study would normally ignore second generation impacts, such as the energy required to fire the bricks used to build the kilns used to manufacture the raw materials.

Most LCAs are simply inventories of materials consumption and releases to the environment: emissions to air and discharges to water and the level of solid waste generated. Carrying out an LCA is a relatively straight -forward exercise, provided the boundary of the study has been clearly defined, the methodology is rigorously applied and the data are accessible.

Authority, to give guidance and to be responsible for the implementation of new purchasing policies. Individual members should be able to provide expertise in matters such as the safety, legal and technical aspects of products. Purchasing officers should have a key role to play in the successful implementation of an environmental purchasing policy. Environmental criteria can be written into purchasing contracts by purchasing officers . In addition to other purchasing controls at their disposal, such as the Trade Act, the Food Safety Act and Consumer Protection Act.

Implementation

The successful implementation of environment friendly purchasing policies depends on education and training to encourage all personnel to think and act critically ion terms of the environmental impacts of products purchased to assess:

- Whether purchases are really necessary – is their purchase essential (rather than borrowing, Sharing or hiring)?
- Whether they are the most suitable for the purpose they are intended;
- Whether their quantities can be reduced:
- Whether they present a hazard to health or the environment

A selection of important elements for the successful implementation of an environmental purchasing policy is presented in Box.

Cost Implications

Specifying environmental criteria in purchasing policies may not necessarily impose extra costs – some comparatively less environmentally damaging products may actually save money (such as certain types of paper or low energy products). There will, however, be other less environmentally damaging products which may cost more than the usual ones. It is therefore recommended that the implementation of an environmental purchasing policy proceeds first with cost saving products. The money saved by these products may then be used to offset the extra cost of purchasing other less environmentally damaging products which do not compete in price with the 'usual' ones.

ECO GREEN CONSUMERISM AND ITS EFFECT ON PURCHASING AND THE ENVIRONMENT

Eco Green consumerism is the basis of environment friendly purchasing policies. This section takes a critical look at the effect of Eco consumerism on Purchasing and the environment and in particular, the

Successful implementation of the purchasing policy

The successful implementation of an environment friendly purchasing policy in the organisation involves:

- Promotion of education and the dissemination of information concerning environment friendly purchasing policy in the organisation.
- Production of publicity concerning environmental purchasing policies and initiatives to inform members of the organisation and the general public;
- Cooperation with unions, local interest groups, other public authorities, environmental groups, Educational organisations and industries-business houses in achieving environmental objectives.
- Consultation with all involved in the organisation's activities during the development and Implementation of environmental policies and initiatives;
- Encouraging the members of the organisation to react appropriately and concertedly to meet the environmental challenges.
- adoption of an environmental policy statement of intent:
- nomination of an environmental contact in the organisation, preferably a member of the management team, supported by key personnel with power to change environmentally damaging activities where necessary;
- identification of principal environmental action areas;
- formulation and implementation of an action plan based on the above principal environmental action areas:
- and monitoring and periodic review of environmental strategy Extent to which eco consumer guides and ecolabels can affect purchasing and environmental impact.

Eco Consumer guides

Of the variety of green consumer guides, one of the most well known is The Green Consumer Guide (Elkington and Hailes, 1988). Others include The Ethical Consumer Research Association's magazine, The Ethical Consumer, which reports on the ethical issues (including environmental concerns) behind brand name products; and The New Consumer organisation's green purchasing guides including Shopping for a Better World (Adams et al, 1991) and The Global Consumer (Wells and Jetter, 1991). All these consumer guides cover a wide range of general purchases. There are also other guides that have been produced for a more specific range and type of purchases, such as The Good Wood Guide (FoE, 1990) or The Universal Green Office Guide (Elkington and Hailes, 1989).

In fact, the Eco Consumer guides have a positive role to play in reducing the environmental impact of purchases. They allow purchasing practice to be aligned so as to minimise environmental impacts, maximise consumers benefit and exert pressure on manufacturers and suppliers to provide goods with a less detrimental environmental impact. It could be argued that additives whether adulteration in food, lead in petrol, phosphates in washing powder, or chlorofluorocarbons (CFCs) in aerosols, are now found less in goods as a result of purchaser and legislative pressure. Using green or eco consumer guides, consumers can encourage and even control manufacturers and/or suppliers to provide goods with less of an environmental impact in two complementary ways: by supporting the manufacturers or suppliers of the least environmentally damaging purchases, and by avoiding the manufacturers or suppliers of the most environmentally damaging purchases. Green Eco consumer guides are also useful for raising awareness about the environmental impacts of purchasing.

On the other hand, however, certain criticisms are levelled at green/eco consumer guides. Firstly, it is argued that the term eco consumerism tends to imply environmental benefit or sustainability, but all manufactured goods (environmentally friendly or not) have an environmental necessity or suitability of products. More often they do not question the need for consumption or purchasing. Secondly, it is argued that eco consumerism has little effect on the way in which goods are produced. The production process is unlikely to be significantly altered as a result of green consumerism, since the consumer is generally less well informed about these processes and less able to impact on the choice of process. Process changes will only occur if industry also becomes environmentally conscious, and/or the cost signals to industry alter. Thirdly, green eco consumerism requires money to change patterns of consumption. Some goods may be less environmentally damaging and less expensive both than their alternatives, in which case a switch in favour of the less damaging goods has both economic and environmental benefits. However, other goods may be less environmentally damaging but more expensive than their alternatives, in which case a switch in favour of the less damaging goods has economic costs. The choice of these goods relies on the willingness to pay for less environmental damage. With limited purchasing power, such a decision is not often easy or obvious. Fourthly, the effect of eco green consumer guides depends very much on accurate and up to date information. With changes in manufacturing processes, for example, such guides may become outdated and obsolete.

In summary, the purchase of all manufactured goods entails environmental impacts, so eco green consumer guides do not identify products without any environmental impacts. The effect of green consumer guides is limited in terms of the production processes of products, and they may be limited in terms of the accuracy and currency of information. The guides have a role in reducing the environmental impact of purchases but can at best only partially reduce environmental impacts, or switch one impact for another.

Ecolabels

A number of ecolabelling schemes have been developed in our country and abroad such as pioneers Canada, the United States, UK and Germany as a way of informing consumers about the environmental impact of products. (One of the oldest and most well established schemes is the German 'Blue Angel' scheme which has been in existence since 1978 and covers around 2000 products (Dadd and Carothers, 1991)

In India ecolabelling scheme was launched more recently in October 1993 following the introduction of the MOERF Regulation on ecolabelling. At the launch, labelling criteria for seven product groups were announced, with criteria for several other product groups promised at a later date: product groups ranging from shampoo to shoes and packaging products to paints and varnishes.

Manufacturers of these products are able to apply for the ecolabel once the relevant environmental criteria for each product group are established.

The Indian labelling scheme takes into account all the environmental impacts of purchases from manufacture through reuse and recycling to final disposal using standardized environmental criteria. The ecolabel can therefore be used to compare the impacts of similar purchases and identify the least environmentally damaging product. It provides information to consumers on the environmental impacts of goods and allows purchasers to make direct comparisons of environmental impacts between similar goods. It also raises organisational and individual awareness about the wider environmental impacts of their purchases.

There are also, however, criticisms of this and other ecolabelling schemes:

(1) Products without ecolabels are not necessarily environmentally worse than those with ecolabels:
They may just have not been submitted for accreditation.

(2) Ecolabels are not a sign of zero environmental impact: manufactured goods, whether accredited by the ecolabelling scheme or not, have an environmental impact (no manufactured product can claim to be free of environmental impacts)

(3) Ecolabelling schemes are limited in their scope. The scheme will never a complete range of product groups.

In summary then, ecolabelling schemes represent a way of assessing the environmental impacts of goods. The schemes offer a comprehensive and standardised approach to labelling and enable purchasers to assess and compare the environmental impacts of similar goods. Purchasers should also, however, the environmental impacts of similar goods. Purchasers should also, however, be aware of the limitations of such schemes. Like green consumer guides, ecolabelling schemes have a role in reducing the environmental impact of purchases but should not been seen as a single solution to the environmental problems associated with the consumption of goods.

PURCHASE AUDIT TERMS OF REFERENCE

The environmentally benign product – an Extremely Rate Beast !

- It is durable and reusable, or recyclable and truly bio-degradable
- It is responsibly and minimally packaged
- It is not obnoxiously frivolous, like a new electric pepper mill
- It releases no persistent toxins into the environment during production, use or disposal
- It is made from recycled material or renewable resources extracted in a way that does not damage the environment
- It includes information on manufacturing, such as location, labour practices, animal testing, and the manufacturer's other business.

Source: adapted from Dadd & Carothers, 1991

The terms of reference of the purchase audit should encompass all the organization's policies and practices that have an influence on purchasing. The terms of reference of the audit should be applicable to all the organisation's different sections or departments. This section describes some commonly used terms of reference for purchase audits. It also summarises the main environmental impacts of a number of different product groups commonly purchased by local authorities.

Products containing paper

Large quantities of paper and paper products are consumed by most organisations. The main environmental advantages of recycling paper and using recycled paper concern issues of energy use and waste disposal. The energy required to produce recycled paper is up to 50 per cent less than that required to produce new paper. Waste disposal and its environmental impact can be relieved by recycling paper and using recycled paper products, thereby reducing the amount of waste paper for disposal.

There are many uses where recycled paper can be substituted for virgin paper. The quality of some recycled paper may be high enough to be used in pricing and reprographic processes. Suitability requires tests to establish information such as dust generation, temperature sensitivity and paper stability (during multiple processes). At the moment the major obstacle for recycled paper use is cost however certain types of recycled paper products which can compete with virgin paper equivalents on a quality and cost basis. These include products such as envelopes and high quality papers. A feasibility study into the use of recycled paper should be made at an early stage of purchasing policy review.

As with all products, the purchase of paper and paper products should be examined such that consumption is minimized wherever possible. This reduces costs and lessens the environmental impact. It will be also necessary to evaluate the appropriate grade of paper to be used for particular uses: certain applications may not require expensive, high grade or pure white papers. Lower grade paper is produced with less environmental impact. Unbleached paper is better for the eye; more eye strain results from reading dark printing on a bright paper background, rather than on a less bright background.

There are a number of ways of reducing the consumption of paper and paper products in organisations. These include the reuse of scrap paper on which only one side has been printed (for jotter pads for example), wipeable notice boards in place of information circulars, reuse of envelopes, double sided rather than single-sided letters and photocopies, and recycling of waste paper.

Energy and products consuming energy

Energy management is a means of saving money and reducing the environmental impact associated with energy production. It is essential that energy demand is minimized if fuel resources are to be conserved and pollution controlled. Current Indian energy generation contributes to approximately 70 per cent of sulphur dioxide and 30 per cent of nitrogen oxide emissions which is responsible for acid rain), and 30 per cent of carbon dioxide, known as greenhouse gas'). It is possible to make energy savings in purchases such as building materials, electrical appliances, heating systems and motor vehicles, etc.

Products containing CFCs and other ozone-depleting chemicals

Chlorofluorocarbons (CFCs) are a group of chemicals which have in the past been widely used in aerosols, refrigerators, solvents, cleaning materials, insulating and structural foams. These chemicals are particularly damaging to the Earth's protective layer of ozone in the upper atmosphere. Eliminating their use and the safe disposal of wastes containing CFCs can help to reduce the impacts of the greenhouse effect.

Products containing lead

Lead is a toxic metal which can accumulate in the environmental and in the body. Small quantities absorbed by the body over time may be sufficient to cause serious health effects. Steps should be taken to reduce the amount of lead in the environment to an absolute minimum. It is also necessary to ensure the safe disposal of wastes containing lead.

Products containing unsustainable materials

The production of certain products involves the use of environmentally unsustainable materials, caused serious environmental destruction and irreversible harm. It is therefore essential that alternatives to such products are purchased. The effects of tropical deforestation, for example, threaten biodiversity and , in some cases, release carbon dioxide stored in organic material with consequent implications for the greenhouse effect. The quarrying of material may involve the removal of natural features from an area as well as noise, vibration and dust nuisances, traffic generation and interference with the water table. The production of peat involves the destruction of scarce wetland habitats.

Products containing asbestos

Asbestos fibres are carcinogenic and thus pose a threat to health. Steps should be taken to reduce the amount of asbestos in the environment to an absolute minimum. It is also necessary to ensure the safe disposal of wastes containing asbestos.

Water and products consuming water

Conservation of water is necessary to avoid depletion of supplies and the environmental effects of lowered water tables in areas of water abstraction. Metered water supplies mean that savings in water consumption can be equated to financial savings.

Products containing bleaches and phosphates

Certain cleaning chemicals may be produced by environmentally unacceptable means or pose a threat to health or the environment in their use. Bleaches and phosphates may have significant environmental effects on water courses. Bleach is an acute poison and phosphates can be responsible for eutrophication. Such materials should be avoided and acceptable alternatives found.

Products containing substances hazardous to health

The Control of substances Hazardous to Health (COSHH) regulations affect the storage and use of certain Office and workplace substances. Many substances including cleaning products, laboratory chemicals, office products, gardening and grounds maintenance chemicals are included in the regulations. If they are labelled as toxic then certain written procedures need to be adhered to. Education and training should be given to those persons working with these substances. It is essential that all persons concerned are made aware of this procedure. Each department or section should have a nominated COSHH coordinator who can be consulted on purchases covered by the COSHH Regulations. Reports suggest that there may be up to 4000 acute poisonings from pesticides each year. Pesticides may have deleterious effects on health and the environment. It is recommended that purchasers seek alternatives to red list and black list pesticides, and avoid wood preservatives containing penta-chlorophenol, lindane and tributylin oxide.

CHECKLIST

Products containing paper

- Whether recycled paper are products purchased whenever practicable?
- Whether waste paper is reused and recycled?
- Whether the appropriate grades of paper purchased according to their use?

Energy and products consuming energy

- Whether low energy light bulbs and tubes purchased?
- Whether energy efficient appliances purchased?
- Whether low fuel consumption vehicles purchased?

Products containing CFCs and other ozone-depleting chemicals

- Whether products made with or containing CFCs (solvents, refrigerants, foams, Aerosols) purchased?
- Whether fire extinguishers containing CFCs have been replaced?
- Whether the use, servicing and disposal of existing products containing CFCs controlled?
- Are non-pressurized pump-action reusable sprays or non-spray products purchased in Preference to pressurised aerosol products?

Products containing lead

- Whether products containing lead (paints, building and plumbing materials, fuel) are purchased?
- Whether lead-free fuel is purchased?
- Whether petrol-fuelled devices have been converted to use lead-free petrol?

Products containing unsustainable materials

- Whether tropical hardwood products are purchased?
- Whether quarried materials are purchased?
- Whether peat is purchased?

Products containing asbestos

- Whether products containing asbestos (insulation and building materials, clutch Linings) are purchased?

Water and Products consuming water

- Whether low water consumption appliances are purchased?

Products containing bleaches and phosphates

- Whether cleaning materials without phosphates or bleaches are purchased?

- Whether environmentally suitable, biodegradable cleaning materials are purchased?

Products containing substances hazardous to health

- Whether purchases for COSHH regulated substances compulsorily ratified by department or section COSHH coordinators?

CHAPTER 12

COMMUNITY AWARENESS

INTRODUCTION

The successful achievement of local government policies and programmes regarding sustainability and environment can only be fully achieved through motivating individual and community action at home and at work. While there are some areas in which local governments can enforce action for example in the field of environmental health there are many aspects where their role is to enable the community. Thus the local authority can encourage recycling through the provision of facilities, or the greater use of public transport and cycling by the introduction of parking restrictions and cycle lanes. However, it may not have any powers of enforcement in these areas. Clearly the success of initiatives will depend on the willingness of individuals to participate. Such participation will in turn depend upon their understanding of the issues and how proposed measures relate to their own environmental priorities in their neighbourhoods and workplaces.

Community awareness is thus important to policy and programme makers who provides the basis to engage people in change. It is important to the community itself, for articulating their needs and priorities and is in better ability to influence decision makeis. The auditor will need to assess measures that are being taken to incorporate community views into policy and programme design as well as actions taken to increase community awareness about issues of sustainable development in the context of Local Agenda 21. This chapter will examine the options open to local authorities to increase community awareness and action and suggest ways in which the auditors can assess the impact and effectiveness. The issues will be addressed under the following sections:

- taking stock of local opinion
- environmental awareness/education
- staff training and involvement
- coordination and partnership

TAKING STOCK OF LOCAL OPINION

Auditors will need to assess the effectiveness of local authority initiatives to incorporate local opinion into policy development and implementation. Several steps will need to be taken in this assessment.

Consultation – with whom?

Local authorities are ostensibly democratic, with local councillors accountable to the electorate, and the first route many people will take to try and influence local programmes will be through their own councillor. However, relying totally on councillors to reflect local opinion is clearly not satisfactory. A relatively small number of people, usually those who are articulate and vociferous, use councillors in this way. Clearly other representative bodies such as resident associations, community associations as well as other voluntary groups, and environmental groups, will need to be targeted as part of any consultation exercise. If the local authority does not itself keep a comprehensive list of such groups one should be available through the local Council for Voluntary Service or Rural Community NGO's. Views of members of the public other than those involved with local community groups also need to be taken into account and mechanisms set up to solicit these. In addition elements of the community which are perhaps not represented by a local group, for example ethnic minorities, the disabled etc., should also be consulted.

Consultation – how?

The mechanisms used to conduct a consultation exercise are numerous and those which are employed clearly depend on its purpose and the budget available. A summary of some of the methods available and their relative advantages/disadvantages is listed in table.

In practice the, best way of eliciting a good cross section of views will be to employ a number of the methods shown in the table. Where resources are limited it may be possible to use students or members of community organisations to conduct local residents' surveys.

CONSULTATIVE COMMITTEES

Having taken account of the public's views in assessing areas of environmental concern in the locality, local authorities will need to address how to ensure that these concerns are translated into policy development and practice. Many LA/councils have, over the last few years established/supported environmental forums or liaison groups at which officers, councillors and selected representatives from local environmental groups meet to discuss LA's/council policies and programmes. Where these do exist they could be given the remit to take on the broader agenda of sustainability. If so it could be argued that their membership should be expanded to include a broader range of community groups to reflect the social welfare aspects of sustainability. However such forums are made up, it is critical that those organisations who are invited to participate feel that their views are really being taken into account and that the spirit of consultation is genuine. In reviewing their effectiveness auditors should be aware that the existence of such consultative bodies does not necessarily mean that community views are being incorporated into policy development. Auditors should be encouraged to question participants, not just council employees, as to the effectiveness of the group.

LIMITATIONS

Devising means of involving the community in the planning and delivery of sustainability plans is obviously crucial to its success. However it must also be recognised that there are limitations inherent in this process as community views are more likely to reflect parochial problems such as litter and dog mess as opposed to factors affecting global warning. There will also be

Undertaking a consultation exercise

Method	*Advantages*	*Disadvantages*
Sample survey of residents	A good cross section of views can be sought Easy to collate response	Time consuming and therefore relatively expensive Style and range of questions may be restrictive
Specific manned contact Point in the authority coupled with local media coverage	Relatively easy to establish and resource Respondents not restricted by scope of a survey form/ questionnaire	Relies on public taking the initiative More difficult to collate responses
Leaflets/questionnaire Distributed through libraries etc.	Easy to administer initiative	Relies on public taking the
Meetings in each area	Allows for full discussion of the tissues	Many people feel inhibited at meetings
Inviting comments from community groups	Easy to administer	Responses don't necessarily reflect concerns of the public at large

conflicts where local concerns are out of tune with the general aims of sustainability, for example public demands for more car parking when LA/Council policy is to restrict town/city centre parking. How these limitations and conflicts are overcome will depend largely on an effective environmental awareness/education programme.

CONSULTATION PROCESS

Local authorities may develop models for 'neighbourhood forums, as a means of decentralising council services and providing a forum for resident groups to articulate local needs and participate in the planning of LAs/Council services. Such initiatives would be well placed to assist in devising methods for the delivery of Local Agenda 21. As the forums are locally based they are more likely to be accessible to local people and their opinions and members could be used to conduct local surveys of residents.

ENVIRONMENTAL AWARENESS/EDUCATION

In order that individuals and communities can develop their own strategies for achieving sustainable development they need to have an understanding of the issues involved and the Council's own policies and planned programmes. In particular people need to recognise what sustainability means to them and their neighbours as well as understanding how this relates to national and international objectives. All too often information about sustainability. Agenda 21, state of the environment reports, local environmental indicators and so on is presented in such a way as they appear remote, full of jargon and an intellectual rather than a practical issue. The marrying of environmental and social welfare goals within the concept of sustainable development should mean a greater relevance to local communities who may well have been put off the environmental movement in the past, as it was perceived as putting greater emphasis on planting trees and recycling than poor quality housing.

This section will look at the scope of local authorities to promote a greater community awareness and the means of achieving this.

Mechanisms of public involvement

	Techniques	*Purpose*	*Resource implication*
1.	Sample survey of residents (door to door)	Discover general attitudes and environmental priorities	Costs significant, but could be marginal if organized/ carried out by local college or university, using students (as in the North Devon audit)
2.	Specific manned contact point in the authority	Giving opportunities for invdividual issues to be raised, and giving a public face to the process	Given the need for at least one dedicated staff member on the audit, the extra costs would be marginal
3.	Local press, radio and TV	Publicizing the issues, increase awareness	Marginal, especially if councillors not staff are interviewed
4.	Leaflets/questionnaires distributed through libraries etc.	Discover general attitudes (but not accurately), and allow issues to be raised	Marginal extra cost
5.	Meetings in each area	Publicity and increased sense of political pressures	Time-intensive for staff. Not justified unless there are specific issues on which to hang the meeting
6.	Request to parish councils for assitance	Specific information; mechanism for getting people involved in the survey work itself	Could extend the scope and value of the audit, at some organizational cost
7.	Invite comments/help from selected list of voluntary groups	Gain extra information, deflect possible suspicion	Very cost effective in terms of increased data and agents of change

Scope

Local authorities may have, through their varied responsibilities, many points of contact with the general public, institutions, community groups and local businesses. Although now diminishing, their role in the provision of education services is still important and one which should be utilised while it is still there. Maximising the potential for developing community awareness through the channels which already exist implies a degree of interdepartmental cooperation and coherence in approach. Auditors should examine whether or not, policies developed in this field are promoted by the LAs/Council as a whole. For example, there may be a programme for increasing awareness about energy efficiency within the home; in this case it would be necessary to ascertain what the Council/LA was doing at its own housing stock and whether information and advice was promoted through residents associations and its own housing office.

Means

There are a number of means through which local authorities can promote environmental awareness:

- Making use of existing publications, newsletters, leaflets, mailings. Thus, leaflets giving dates for Bank holiday arrangements for the refuse collection may also include details of local recycling facilities. Another example might be distributing information about car-sharing together with pay slips. The auditor should investigate whether all Council publications are being used effectively to promote awareness of sustainability.
- Maximising the use of the media, not only the local press, radio and television but also community newsletters etc.
- Events – for example, exhibitions, touring, local libraries and other estates, public meetings etc.
- Major projects such as environmental centres discussed in succeeding paragraphs.
- Working with existing environmental and community groups – these may be more effective than the LA/council itself in raising community awareness of the issues. Groups such as environmental NGOs, local Wildlife Trusts etc. have for a number of years produced information and advice on

environmental issues. The Council could use its powers as a grant-giving institution to fund existing groups with particular skills in this area to run awareness-raising campaigns.

- Developing programmes aimed at reducing a household's impact on the environment. These would include basic information about what householders can do to reduce waste, improve energy efficiency, minimize car use etc. with follow up questionnaires to ascertain whether participants have pursued any of the options open to them.
- Working with the private sector, through bodies such as the Chamber of Commerce to promote Environmental awareness in the business sector.
- Working with schools to build in environmental issues into cross-curriculum activities; encouraging schools to undertake practical environmental projects, for example energy efficiency programmes, a wildlife area, etc.

It will be important when deciding avenues to pursue that local authorities make assessments as to what is the most cost-effective means of raising awareness. It may be that launching a major project, without private sector support, may be less cost effective and certainly involve more risk than working through existing structures and community and business links, to promote a greater awareness of the issues.

It is also important for auditors to make some assessment of the accessibility of the LAs/Council's publicity material. At its most fundamental level it is clearly important that in areas where Hindi or English is not the first language of people within the community local language literature can be made available in all appropriate languages. It should also be accessible in style and give examples as to how the issues discussed are relevant to that particular community.

EXAMPLES OF AWARENESS-RAISING INITIATIVES

Major Project

The Environmental Action, Training Education project may be developed by environmentalists groups of nature case. The aim of the project will be to provide a stimulating and imaginative environment where people living or working in surroundings and visitors to the city as well can experience and learn about a variety of environmental issues and ideas. The action group may have three functions:

(1) *Visitor Centre :* which may provide easily accessible information on general and specific environmental issues to the general public. The centre will be educational, experimental, interesting and above all a fun.

(2) *Training and Education Centre:* which may provide a range of flexible spaces designed to facilitate the promotion of detailed awareness of environmental issues to individuals and organizations. These will include a "single window" business advice centre and environmental information facility.

(3) *Accommodation:* for appropriate environmentally sensitive organisations.

While the project is led by local authority, the concept has been developed in close collaboration with the community and business sectors.

CASE STUDY : NETHERLANDS

Programme to encourage greater awareness of and participation in environmental improvement measures – Richmond Ecofeedback Scheme

Ecofeedback began in the Netherlands in 1979 when 18 households took part in a scheme to reduce their energy consumption. House holders were given a target for energy reduction and were asked to report progress to the authorities running the scheme. The scheme has spread and now more than one quarter of the households in the Netherlands take part, reducing gas consumption by an average of 10 percent. The London Borough of Richmond was the first local authority in the UK to take the principle of Ecofeedback and apply it to waste minimization. The scheme was launched during the 1992 National 'Watch Your Waste' week and participants were given information about ways of reducing their waste and asked to record their weekly waste arisings. The majority of participants reduced their waste by over 10 per cent.

STAFF TRAINING AND INVOLVEMENT

It is clearly important that if local authority staff are to be responsible for policy development, evaluation and monitoring that they themselves should have a firm understanding of the ensure resistant and sustainable issues. Auditors should examine staff training programmes to assess whether this matter is being addressed. There are often within local authorities pockets of specialised knowledge and experience and where this is found, for example in energy efficiency measures, such staff should be involved in devising training for other staff. It may be that people with suitable skills exist outside of the local authority, in the local community, who could provide training on particular issues. For example, many grassroots environmental group/NGOs may have built up expertise in certain areas which could be utilised in Council/LAs environmental training programmes.

It is clear that sustainable development cannot be achieved by local authorities exercising in isolation. It is thus important that local authority staff are given appropriate training in working effectively with other bodies.

While this section has focused on local authority staff, there is also a need for councillors to be better informed about the issues surrounding sustainable development. Councillors are themselves supposed to be a very busy people and need concise and relevant information upon with they can make decisions.

COORDINATION AND PARTNERSHIPS

Coordination within local authorities

Coordination across departments of LAs/council programmes and policies for sustainable development is clearly important if progress is to made. This is not always easy: while environmental issues may be high on the agenda in Planning and Environmental Health departments, this will not always be the case in other departments. Clearly councils need to adopt a corporate view on the priority given to issues of sustainability and develop a strategy for progressing programmes across the LA/council as a whole. In some areas this has been achieved by creating a new committee, reporting to the main policy and resources committee, drawing representatives from each department to progress the council's environmental policies and programmes.

Partnership with private/community sector

Sustainability can only be achieved through cooperation between the three sectors involved public, community, private. In their role as enablers it is often the local authorities who take the lead in developing partnerships as a means of effecting change. Successful partnerships need all parties to recognise their relative strengths and weaknesses, to make the best use of existing resources and thus avoid duplication of effort. Successful partnerships, especially across different sectors, each with their own cultures require commitment and time., If they are to reach their full potential. Auditors should examine the ways in which local authorities are working with other sectors and assess how effective such partnerships have been in initiating local programmes for change. As more and more LA/council services are contracted out to the private sector, it is clearly important that as far as possible contract specifications take account of measures to further sustainable development.

CONCLUSION

Community awareness is arguably the most important area to be tackled if sustainable development is to be achieved. Local environmental groups such as Nature Care organisations, Wildlife Trusts and Energy Action groups can be a significant deal to promote community awareness of environmental problems and promote practical solutions. Since environmental issues have become mainstream, and have through the concept of sustainable development been broadened to include social objectives, it is now the responsibility of all sectors and interest groups to become involved in the process of promoting awareness. Local authorities during awareness need to be imaginative in their approach and show a willingness to work closely with others in order to achieve their objectives in this field.

CHECKLISTS

The Consultation process – Involving the public

- Whether the LA/council already have contacts with community groups?

(a) Is this list comprehensive?
(b) How are these contacts spread across the various departments?
(c) Is there any coordination between different departments regarding community- contact?

- Whether appropriate mechanisms have been used to elicit local opinion?
  How effective have these been in terms of

(a) response?
(b) Ensuring local views are incorporated into policy/programme development?

- Whether during policy/programme development plan, local views sought at different Levels/Stages.
- Whether local community groups/environmental groups and their specialist knowledge have been made in use of
- Whether the LA/Council has established any liaison group?

(a) How is the membership of such groups drawn up?
(b) Is the membership representative of the various interest groups?
(c) How effective is the group
  - From the Council's point of view?
  - From the participant's point of view?

- Whether LA/Council services are organised – centralised or decentralised?
- Whether there is scope for decentralising services to encourage greater community Participation in planning and development?

The Local Authority role in environmental awareness

- Whether the council have a coherent corporate approach to promote environmental awareness?
  Who is responsible for developing a strategy?
  Do they have access to facilities across all departments?
- Whether there was a budget available?
  Who is responsible for it?
- Whether required means have been employed by the Council to promote environmental awareness?
  Has the Council investigated all available options?
- Does the Council work with existing community and environmental groups to Promote awareness?

- Is care taken to make publicity material relevant to the particular local situation, use of Appropriate language, illustrative local examples etc.?

Development expertise – training for staff and councillors

- Whether awareness of sustainability was built into staff training? How is this done?
  Does it maximize the use of expertise within and outside the Council?
- Whether training provided on the concept and practicalities of developing partnerships With outside agencies?
- Whether local authorities/councillors aware of the issues of sustainable development?
  Have steps been taken to inform them?
  How successful have these been?

The Corporate approach to sustainability

- Whether the LA/Council aware of the importance of developing partnerships?
- Whether there is a corporate view on how such partnerships should be developed or has the lead been taken by individual departments?
- Whether there is a high level of cooperation between departments towards achieving sustainable development?
- Whether Council had the experience of developing/taking part in partnership projects?
  How successful have these been?
- Whether Council services have been contracted out, has the contract specification taken into account environmental objectives?

RIO DECLARATION ON THE SYSTEM OF PUBLIC PARTICIPATION

Rio Declaration on Environment and Development states ABOUT THE PEOPLES PARTICIPATION that "Environmental issues are best handled with the participation of all concerned citizens, at the relevant levels".

In order to increase awareness of environmental problems and promote effective public participation, access to environmental information should be guaranteed.

Public participation contributes to the endeavours of public authorities to protect the environment, and bearing in mind that environmental policy and decision making should not be restricted to the concerns of authorities.

In order to promote effective public participation the public needs to be aware of the means and methods of participation in environmental decision making processes, and in the solving of the environmental problems.

Public participation can be a source of additional information and scientific and technical knowledge to the decision makers.

Environmental authorities should raise public awareness in order to promote greater public understanding and support for environmental policies and its enforcement.

The promotion of public participation requires the transparency and the accountability of public authorities, thus improving their credibility and strengthening support for their activities.

When emphasising the importance of public participation in protecting environmental rights, it should also be recognised that all persons, both individually and in association with others, have a duty to protect and preserve the environment.

Whereas practicable access to the courts and administrative complaints procedures for individuals and public interest groups will ensure that their legitimate interests are protected and that prescribed environmental measures are effectively enforced and illegal practices stopped.

I. ACCESS TO ENVIRONMENTAL INFORMATION

1. For the purpose of these guidelines, environmental information means any information on the state of water. Air, fauna, flora, land, natural sites, and on activities or measures adversely affecting or likely to affect these, including administrative measures and environment management programmes.
2. Any natural or legal person should have free access to environmental information at their request, subject to the terms and conditions contained in these guidelines, without regard to citizenship, nationality or domicile and without having to prove a legal or other interest.
3. Public authorities (at national, regional and local level) and bodies having public responsibilities for the environment, with the exception of bodies which are acting in a judicial or legislative capacity, should supply environmental information subject to the terms and conditions contained in these Guidelines.
4. Public authorities should regularly collect and update adequate environmental information. In addition, States should establish, where voluntary systems are inadequate, mandatory systems for ensuring that there is an adequate flow of information about activities significantly affecting the environment to the public authorities.
5. States should take the necessary steps to make their environmental information systems more transparent, e.g. by specifying the type and scope of the environmental information available and the basic terms and conditions under which it is made available and the process by which it can be obtained, and by the establishment and maintenance of registers and the designation of information officers.
6. A request for information may be refused only where it affects:
   (a) The confidentiality of the proceedings of public authorities, international relations and national defence;
   (b) Public security;
   (c) Matters which are, or have been, *sub-judice* or under enquiry (including disciplinary enquiries), or which are the subject of preliminary investigation proceedings;

(*d*) Commercial and industrial confidentiality (for example in relation to agricultural and other business activities), including intellectual property;

(e) The confidentiality of personal data and or files;

(f) Material supplied by a third party without that party being under, or being capable of being put under, a legal obligation to do so, and where that party has not consented to the release of the material;

(*g*) Material, the disclosure of which could endanger the (environment, e.g. information on the breeding sites of rare species.

A request may also be refused if it would involve the supply of any material in the course of completion. The aforementioned grounds for refusal are to be interpreted in a restrictive way with the public interest served by disclosure weighed against the interests of non-disclosure in each case. Reasons for a refusal to comply with a request for information must be stated in writing. Where only part of the information requested falls within one of the exempt categories, the remainder of the information should be separated out and supplied to the person making the request.

7. Public authorities should respond to a person requesting information as soon as possible and at the latest within six weeks.

8. Environmental information, such as that contained in public registers, should be available to the public for inspection free of charge. Any person requesting information should be provided with adequate facilities' for obtaining copies of such information (subject to copyright provisions) on payment of cost of reproduction and dissemination, if appropriate. Where, information is held in various forms, it should be provided in the form specified by the person requesting the information, e.g. in written, visual, or electronic form.

9. States should ensure that a person who considers that his or her request for information has been wrongfully refused or ignored, or has been inadequately answered by a public authority, or overcharged, may seek judicial or administrative review in accordance with the relevant national legal system.

10. States should regularly publish up-to-date information on the state of the environment, e.g. in a report.
11. States should actively publicize the availability of important national and international documents on the environment, where they exist such as strategies, programmes, action plans and progress reports on their implementation.
12. States should actively publicize the availability of the texts international legal instruments, to which they are a party, and which establish procedures for public access to environmental information or public participation rights, preferably in their own language(s) together with relevant conference resolutions or recommendations.
13. States should inform the public of the possibilities of submitting information to international bodies concerning non-compliance with international rules.
14. States should encourage entities whose activities have a significant adverse impact on the environment to report regularly to the public on the environmental impact of their activities.
15. Public access to information stemming from such voluntary schemes as eco-audits should be encouraged, as should eco-labelling schemes for more environmentally friendly products.

II. PUBLIC PARTICIPATION

16. States should facilitate public participation in environments decision-making processes and decision-making processes having significant environmental implications.
17. States are encouraged to establish formal and informal consultative processes to facilitate the involvement of NGOs in decision-making processes having significant environmental implications and to eliminate impediments or obstacles to public participation.
18. States should make special efforts to promote public participation in environmental policy-making and decisions that are of particular interest to regional and local communities.
19. Consultations should take place early in the decision-making process, at a stage when options are still open and effective public influence can be exerted. States should establish transparent procedures and providing relevant information.

Where appropriate, the relevant authorities should give the public additional assistance and explanations. States are encouraged where feasible, to relate time limits placed on public consultation to those under the access-to-information regimes with a view to ensuring informed public participation

20. The relevant authorities should be responsible for the effective training of public officials to improve their understanding of their responsibilities in granting the public access to information and facilitating public participation in environmental decision-making.
21. Before decisions significantly affecting the environment are taken, States should introduce measures ensuring that public opinion, including the views of NGOs, other interest groups and environmental advisory bodies, is taken into account.
22. States should ensure public participation in environment administrative decision-making processes preferably by means of explicit rules governing certain procedures such as, if applicable, environmental impact assessment (EIA) and the issuing of permits or licences, particularly where these may have significant effect on the environment. Those rules could include, *inter alia,* the right to be heard, procedures which include the right to propose alternatives where feasible, a reasonable time to comment, the right to a reasoned decision and the right of recourse to administrative and/or judicial proceedings in order to challenge failures to act and to appeal decisions.
23. States are encouraged to take as a minimum standard the obligations and recommendations on EIA as contained for example in the Convention on Environmental Impact Assessment in a Transboundary Context (Espoo, 1991).
24. States should ensure that persons involved in public participation in environmental matters are not penalized in any way for activities that are otherwise lawful.

III. ADMINISTRATIVE AND JUDICIAL PROCEEDINGS

25. The public should have access to administrative and judicial proceedings, as appropriate. Suitable legal guarantees should ensure that proceedings are fair, open, transparent and

equitable. It is desirable that proceedings are not prohibitively expensive.

26. It is desirable that standing should be given a wide interpretation in proceedings involving environmental issues.

IV. IMPLEMENTATION OF THE GUIDELINES

27. States are encouraged to adopt necessary strategies for the implementation of the present Guidelines, which should be developed as a result of a broad consultative process.

28. The effective implementation of access to environmental information and public participation in environmental decision-making processes calls for the establishment of a clear regulatory framework providing procedural and institutional guarantees and proper enforcement programmes. Where appropriate. States should set up organizational structures to facilitate the effective operation of the above guarantees, e.g. designation of information officials and officials to promote contacts with the public, the allocation of environmental responsibilities to an ombudsman, etc.

29. States should recognize the special role of local and regional governments and delegate the necessary authority to these bodies to ensure implementation of these Guidelines.

30. States should promote environmental education and training for the general public and specified target groups, especially regarding the methods and techniques of access to information and public participation. The decisive role of NGOs, educational institutions and the media should be recognized and they should be given appropriate support.

31. States should promote regular monitoring of the implementation of the present Guidelines. States are requested to support ongoing activities and facilitate exchange of experiences of implementation. States should report about the progress made in implementing the present Guidelines to the United Nations Economic Commission for Europe not later than two years after the adoption of the document.

Part-Two

ANALYSIS OF CURRENT PRACTICE

CHAPTER 13

DEVELOPING AN AUDITING STRATEGY

INTRODUCTION

This chapter focuses on the technical tasks of developing an EA strategy. The discussion starts by looking at the overall makeup of a strategy and then examines specific stages which can help in developing a coherent approach.

The chapter has three main parts. The first section examines the background and evolution of approaches to EA. The middle section then builds on this by identifying the different tasks involved in EA and how these may be combined into a strategic process. The aim is to arrive at a coherent and manageable framework for achieving quick but lasting process. The final sections elaborate on the specific requirements of each of the stages in the strategic process.

BACKGROUND OF AN AUDITING STRATEGY

One of the most important tasks of the survey was to find out what types of audits were being undertaken by authorities. There is now a much clearer picture of what constitutes EA, whereas at the time of the survey the purpose and extent of the processes involved was not self-evident, even for many of those involved. Earlier the types of audits, with the major distinction being between internal auditing (IA) and state of the environment (SoE) reporting was discussed. A process combining IA and SoE reporting is referred to as a full audit e.g. state of the environment report and Internal Audit. Although these distinctions are now becoming up dated and the elements of what may be considered to be an EA process have evolved, the basic components can still be viewed as internally focused (IA) or externally focussed. This applies particularly where issues of scope are being discussed.

There is a clear preference for IA over SoE reporting. This conforms to most of the advice given by the local authority associations and is also in keeping with the most recent developments, specifically with respect to environmental management systems. The emphasis is very much on putting one's own house in order before spending time and money examining the detailed nature of the environmental problems. There is a very pragmatic logic behind this approach, which is undeniably attractive in these times of financial restraint. It is also supported by the imperative to begin reducing unsustainable behaviour even though the exact direction of sustainable development policies has not yet been worked out. So, the evidence is that given a choice between internal and external auditing a local authority/company is most likely to go down the path of IA.

Combining the auditing tasks

An interesting finding is the number of authorities who has been identified for conducting a full audit. Once again, there are clear advantages from pursuing this combined approach. The data from the SoE can help inform policies and practices in the IA and these can then be subsequently monitored and their progress assessed by referring back to the quality of the environment data.

Type of Audit

The important thing is not to get tied up in terminology and typology. If the basic purpose of an SoE is seen to be gathering information on the state of health of environmental features, then it follows that all audits will to some extent involve an SoE component. The issue then becomes one of degree. The amount and focus of information collection will vary depending on the tasks for which it is intended. This can be most clearly illustrated by discussing the various tasks within the context of an overall environmental strategy. Figure Approach A and Approach B presents two quite standardized approaches to developing an EA strategy.

Approach A is based on the premise, as outlined above. It is better to being the process of change as soon as possible. The most accessible issues and probably the most amenable to change are therefore those related to internal practices and this would include management systems also. These practices are intimately linked to the authority policies, so once the process of practice review has begun the audit will inevitably also draw in some policy areas. For instance, aiming to raise internal environmental awareness should lead to the generation of ideas and the production of materials which can then be used in a community awareness programme. Another examples is where changes in internal purchasing arrangements can result in amendments to contracting policies (to include environmental criteria) and perhaps even changes in investment strategy (stressing ethical and environmental provisions).

The product of the internal audit will be some sort of action plan, which sets out for internal and/or external

APPROACH A	**APPROACH B**
Statement of	Statement of
Principles	Principles
RIP	SoE
PIA	RIP
Action Plan	PIA
SoE	Action Plan

Basic approaches to EA

Consumption what the authority intends to do as a result of the audit findings. However, somewhere along the way it should become clear that assessment and decision making could only ever is limited unless more information is brought into the system. This realisation might occur while still in the process of producing

Table 13.1 : Matrix recommended for plan appraisal

	Global sustainability					*Natural resources*					*Local environmental quality*				
Criteria	*1*	*2*	*3*	*4*	*5*	*6*	*7*	*8*	*9*	*10*	*11*	*12*	*13*	*14*	*15*
Policies	Transport energy : Efficiency : trips	Transport energy : Efficiency : models	Built environment Energy: efficiency	Renewable energy potential	Rate of CO 'fixing'	Wildlife habitats	Air quality	Water conservation and quality	Land and soil quality	Minerals conservtion	Landscape and open land	Urban environmental 'liveability'	Cultural heritage	Public access open space	Building quality

Suggested impact symbols	No relationship or insignificant impact	Likely, but unpredictable impact	Significant beneficial impact
	Significant adverse impact	Uncertainly of prediction or knowledge	

Source : DoE (1993) Environmental appraisal of development plans.

the first action plan or perhaps during a subsequent review. It will certainly soon be apparent that only a very limited system of monitoring will be possible without some sort of SoE process. Judging the effectiveness of many of the policy recommendations. For example, in the area of pollution control or awareness raising, this will involve environmental surveys.

Approach B, through less often had chosen, has an equally sound logical basis. For some authorities, it may be considered to be little point in assessing policies and practices and recommending a set of actions without first taking stock of the detailed nature of the issues being addressed. Most authorities recognise that although they have a direct influence within their region, for example as a resource consumer and potential polluter, they have an even greater indirect role to protect the quality of life of the region's inhabitants. This reflects the often-vaunted position of local authorities as guardians of the local environment. Authorities pursuing this approach would therefore conduct some sort of SoE survey either before or in conjunction with an IA and would use all this information to produce their action plan.

These approaches, though representing divergent positions, can be equally successful. They have been presented as processes, which happen within the bubble of an enclosed local authority. This is clearly not a realistic scenario. Many external factors will impact on every stage of the process. The principles will, for instance, have been framed by officers and members and will be based on their perceptions of the environment. These perceptions will naturally be greatly influenced by the quality of the local environment as they see it themselves and as it is portrayed by various interest groups. Similarly, the RIP and PIA each address issues which on one level are internal to the authority but are ultimately about the authority's impact on the wider environment. Both the purpose and consequences of the internal assessment are therefore inextricably linked to a consideration of this wider environment. In the same way, no SoE can be conducted in isolation from its policy context. They may be most clearly evident in the standards by which the quality of the environment is judged in terms of legislation and local authority policies. The SoE will at the very least require an evaluation of the environment in relation to these standards and the results of the evaluation will be a series of recommendations for what the authority, among

Table 13.2 : Stated objectives of the Local Areas Environment Audit

- Enabling the authority to assess the environmental impact of their policies and practices.
- Producing an action plan, charter or strategy for the enrvironment.
- Providing a baseline of environmental data in order to monitor progress and change
- Helping the authority to play an enhance role in resolving environmental problems.
- Enabling the authority to raise the profile of environmental issues.
- Heightening awareness about the condition and needs of the environment.
- Improving the image of the authority as an environmental role-model
- Identifying gaps in environmetal data
- Providing information to assist the authority in their enabling/ advisory role.
- Increasing public access to environmental information.
- Encouraging and coordinating cooperation between the many agencies - public, private and voluntary - which have an environmental role.
- Widening public participation in environmental decision making.
- Providing a resource for educational use.
- Engendering the concept that environmnetal quality and economic well-being are complementary.
- Other

others, needs to do to improve matters. Therefore, it is clear that the terms of reference and the output of the SoE link directly to the IA.

These points should serve to illustrate that eventually all audits combine elements of both IA and SoE. The difference between the approaches is in timing and emphasis. Evidence for this appears in the discussion which follows.

Taking the process forward

When one thinks about the range of issues involved - the number of practices and policies that impinge on the environment, the interlinkages between all of these, the amount of change required within the organisation to take these into account, the complex web of social, economic as well as environmental factors, etc. - then the logic of a staged approach becomes unavoidable. Apart from the scope of issues, EA in order to operate as an ongoing process, it must have a system of feedback and review whereby success can be evaluated and new issues brought on board.

A SUCCESSFUL AUDITING STRATEGY

A strategy is largely about getting the right elements into the right order. The difficulty is in knowing what is 'right' for each particular circumstance, whereas a clearly identifiable set of procedures with formal decision points may suit one authority, it may be totally inappropriate in the context of an authority exploring alternative processes or seeking to accommodate wide-ranging involvement. The discussion in the remainder of this chapter proposes a set of elements which, when combined, represent a strategic framework for EA. The recommendations that are made are not intended to be prescriptions suited to all authorities but rather suggestions which will hopefully provide a starting point for developing a strategy suitable to its particular setting.

The framework proposed draws on three main factors:

(1) Over the last few years a clear pattern has emerged from experience of the elements that should be in an auditing strategy if it is to have a good chance of success.

(2) Many of these elements have been crystallized in the EMAS scheme, which presents an excellent Framework approach likely to become the standard against which audits are judged in the future.

(3) *Over and above everything else, what really matters in a successful audit is the degree of commitment there is to bring about real change. A robust strategy can aid this commitment but it can not replace it. So to a degree, any strategy that suits the authority and focuses on engendering commitment has a good chance of success.*

Initial thinking

Earlier the two approaches to EA were discussed. The features from these approaches are recreated here below, but this time the emphasis is on a strategic process which involves inter linked activities and feedback.

Simplified as this model is, it provides a reasonable representation of the stages in the process and is generally applicable to all audits. IT is just this type of model that has informed the initial thinking of many authorities engaged in EA. A more detailed analysis of the stages of the process will follow, but first the relationship between the main assessment elements (RIP, MA, PIA, SoE) is discussed.

Prioritizing the tasks

It was noted in the previous section that all audits combine these assessment elements to varying degrees and to do so, while still maintaining a manageable process, necessitates adopting a staged approach. This in turn requires a strategy for ensuring that continual progress is actually achieved and managed within the authority, while at the same time an overview of the entire process is maintained. The key to this is to focus the objectives of the audit so that the emphasis is always on the design and implementation of actions. This might sound like an obvious point but it is actually at the core of the problems encountered in a number of authorities.

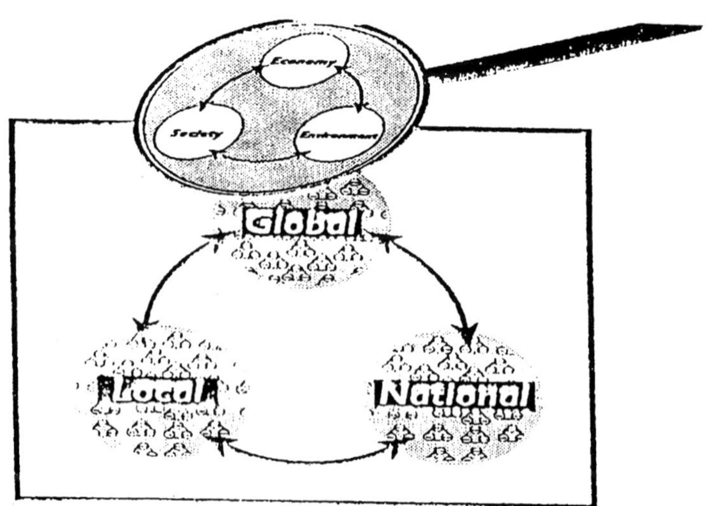

A systems view of environmental concerns

UNFORGETTABLE ENVIRONMENT EVENTS IN AUDIT THE IMPLICATIONS OF NOT KEEPING THE AUDIT ACTION-ORIENTATION

- 'West Country' CC was one of the first countries to get involved in a detailed audit, with the focus predominantly on internal practices and some areas of policy. Much of the analysis was very perceptive and the process was conducted in a reasonably open and participatory fashion. Although the final report contained some recommendations for action, these were not worked out very comprehensively and the action phase was in many ways considered as a separate process. The result was than an action plan did not emerge until well over a year later, by which time much of the impetus had faded. In hindsight, the initial audit assessment might have been over ambitious or perhaps the process was not managed well. What is certain is that the implementation or output phase of the process received too little attention early on and the audit was not driven by the imperative of action.
- 'East Country' DC was also one of the first authorities to see the potential of EA, this time with a focus on SoE reporting as a stage before IA. Consultants were hired to produce the initial audit report and, although it took much longer than originally expected, the results were very encouraging. However, while the scope for the audit spanned the whole gambit of internal and external issues, it did not have a specific brief to produce a comprehensive set of recommendations. This was to be a follow-on task for an internal steering group. For various reasons the action proposals did not emerge for nearly a year and when they did it was not a detailed recommendations but as compromise, no cost, solutions. The process had to all intents and purposes broken down and despite an impressive assessment of issues the auditing effort bore little fruit.

Listed below are a number of suggestions to help achieve this action orientation:

Avoid trying to be too comprehensive. Accept that the issues can be extremely complex and will require detailed assessment. Therefore, narrow down the scope right from the start.

Focus on those issues where there is the greatest opportunity to bring about change. The issues of relevance will vary for every authority, depending on location, tier of government, knowledge, expertise etc. This does not mean avoiding. SoE issues but it does mean that any environmental assessment must be tied in with internal mechanisms.

Remember that successful change is as much about manageability as about direction. There is little point in producing a strategy for sustainable development, for example, if the institutional mechanisms and resources are not capable of delivering it.

Don't allow the process to become top heavy. Defining principles, setting aims etc. can be very difficult and engrossing tasks. The time spent on them should initially be deliberately limited to allow progression to assessment and action. Later in the audit cycle, principles and aims can be returned to and if necessary refined.

If these few suggestions are taken on board the priorities for the audit should become much clearer. The exact sequence of issues to be addressed is largely a matter of choice for those directing the process. In deciding they should have regard to the views and aspirations of the Council/LA personnel and of the community (through perhaps a public opinion survey and a forum of interested parties). Mechanisms for wider involved will be addressed the succeeding chapter.

The final point to make in relation to developing a strategic EA process is that it must be manageable on a corporate level. Mechanisms to achieve this will be discussed later but from the outset the process should be designed with this overview in mind.

On a practical level this will be facilitated by a clear statement of principles and aims but also by explicitly addressing the interrelationships between issues and levels of impact. Another way to achieve this to resolve, right from the beginning of the process, to consider the environmental impact of all-new policies and council decisions, perhaps as a paragraph in committee reports. This relatively simple act of notification could have tremendous benefits in terms of raising awareness and ensuring that the audit does not become a dislocated or sectoral initiative but is amalgamated within the entire workings of the authority.

TOWARDS AN ORGANIC MODEL OF THE AUDITING PROCESS

The framework presented below is a development of the basic strategic process outlined in the previous section. The same elements are there but now they are amalgamated into a unified representation of the auditing process. It is a reformulated version of a basic rational (or normative) approach to planning and decision-making. Here lie its strengths and its potential weaknesses. It is this type of approach that has informed much of the plan formation and policy development in the last few decades. It facilitates logical analysis, sequential progress and should provide clear information to aid decision-making and guide action implementation. This, at least, is the theory! The reality is naturally a bit messier. Without entering into the larger debate, suffice it to say that in practice the conditions of the rational process are rarely fully met. Analysis is only ever partial progress occurs more cyclically and erratically then can be planned for and, most importantly, decisions are not based on clear and full knowledge. These weaknesses in the rational approach suggest three ameliorative solutions:

(1) An attempt should be made to amend the model to take account of these factors.

(2) The elements in a model should only ever be adopted if it suits the interests of those using it to do so. Flexibility and adaptability are essential, so a model must only be a starting point for a process and not viewed as a blueprint.

(3) The policy process, as represented by the model, should be recognised for what it is - simply an aid to help those who are actually managing the process. Another focus of attention should therefore be the organisation, management mechanisms and individuals involved in EA.

It is around the first two points that the remainder of this chapter is presented. It will examine each stage in the auditing model to seek a rational but realistic explanation for how to proceed. The approach to doing this has already been largely outlined in the introduction. Basically, it involves looking at past experience and combining the findings of what has happened with what could and should happen, if the conditions are right. There is a hard use that these recommendations will come across as prescriptive statements but they are not meant to be viewed in that way. Figure (below) simply represents a set of logical stages while the discussion of each of these stages tries to produce some informed advice. Limitations are highlighted but the emphasis is on practical steps for making progress with EA.

Following this, the next chapter will present and evaluate a management framework that takes account of some of the realities of policy development and tries to direct these towards producing required change in the right direction.

DEFINE PRINCIPLES AND SET AIMS

Often, it is considered as a rather vague and slightly abstracted stage from the rest of the process. Principles and aims are often set out at an early stage. Sometimes in a separate 'Environmental Policy Statement' or 'Green Charter'. As such they obviously represent some of the initial thinking on EA and cannot be expected to be a totally accurate representation of what will follow. This does not necessarily matter. There will be plenty of time later in the process, or in a subsequent review, to amend and update these initial statements. This does not, however, mean that how they are set out originally is unimportant. Their influence lies not in their exactness but in the expression of commitment. In fact, they are the very foundation stone for the entire process. The following important points

- **It represents the authority's commitment to act** They incorporate the inspiration for the auditing activities and the set of beliefs around which these activities are developed. As such they will largely determine the type of audit, the scope, the resources made available and the sort of output expected. In other words, they characterize the entire auditing process.

- **It defines the direction the audit should take**
  In particular, the aims are the starting point for all the assessment and review stages which follow. It is against these that the audit will ultimately be judged.

- **It provides the initial strategic framework**
  A working framework for all the other stages in the process and for management is often developed at this time. While this should be flexible and adaptive, it will certainly be a major influence on what follows.

A discussion on the importance of commitment and of ways to help engender it forms a major part of next chapter. This is because commitment is to a great extent an issue of individual beliefs and perceptions. While framing the statement of principles and aims this must be kept in mind. Apart from this, the only advice that can be offered is that the statements should be as open and honest as possible and that they should result from a truly in-depth exploration of what is hoped for from the process.

In the preceding paragraph/ we have listed a sample of the sort of principles and aims that often appear in audits. The key phrases, as such, are:

- Local action
- Participation
- Environmental protection and enhancement
- Institutional change
- Sustainable development
- Balance and harmony

It gives a reasonable indication of what EA is all about.

A final point to make about this stage is that the thorny issue of resources should be squarely faced. One of the greatest determinants of the type of process that will evolve is the amount of resources (both time and money) that can be made available. There is no point in ignoring this issue until say, the decision making stage because much of the assessment will require a knowledge of the potential resource input (refer next chapter)

ESTABLISH GENERAL OBJECTIVES

Principles and aims do not in themselves provide sufficient direction for what can be a complex and multi-levelled process.

They must be taken and are refashioned into a coherent and pragmatic set of objectives. At this stage the objectives should still be quite general but must focus the aspiration in such a way that the aims are kept in sight while the process of change is kept manageable.

The stated objectives illustrate in descending order of importance the general objectives based on categories suggested by the 1990 'Environmental practice in Local Government' guide, UK, which may be accepted in our country .

A number of key issues arise from an examination of this data:

(1) The majority of authorities may be endorsed at least nine objectives. It makes clear that moat authorities have a range of quite broad and diverse objectives for their EA process. It could also suggest a certain vagueness as to what exactly is be achieved, the audit perhaps being viewed as a potential panacea for a wide variety of environmental ills

(2) It suggests a laudably pragmatic approach. Evolutionary nature of the process and the lack of experience. With handling the type of change involved most authorities are initially quite tentative and concentrate on the project itself rather than some of the longer term aims.

(3) Most fundamentally, there appear to be two levels of EA active at the same time. The first is an inductive process, which is tuned to gather information in order to formulate strategies and define aims. This is suggested by the top three objectives (in Figure), which shows that audit to be a first attempt at a comprehensive and corporate environmental strategy for the vast majority of councils. The second level of EA is that of normative process. As suggested by the theory, it seeks to satisfy prescribed aims and to guide actions towards their fulfillment. The Figure shows that the pursuit of stated principles and aims (participation, environmental enhancement, resource conservation etc.) is generally given a low priority.

The contradictions raised by this dual purpose are quite fundamental not least in deciding towards objective efforts and resources should be directed. The solution lies, as suggested earlier in the development of a strategic process. Within this process the two levels become eminently compatible. The objectives can be pursued in stages, with feedback from the assessment and review directing the priority to be assigned to each. In this way, establishing a firm auditing process will facilitate some immediate though probably short-term action while at the same time paving the way the more long term changes which will be required.

INITIAL OPERATIONAL OBJECTIVES

The term operational objectives are some of the most important components of an audit. This refers specifically to the tasks of scoping, setting targets and identifying suitable indicators. These tasks represent the worked-through details of what is to be achieved in the EA process.

Scooping

Defining the scope of the audit has already been discussed in a number of respects, particularly in relation to how comprehensive or partial the initial approach should be (earlier discussed). Taking on board the recommendation that a phases and strategic process is the most appropriate way forward, the issue then becomes one of prioritizing areas of concern. The following three criteria can be employed:

- **Scale of concern** : A local authority will inevitably focus on issues of local concern, but should do so with an eye to other scales of impact. It is meant by the concept Think Globally, Act Locally' - local change is part of a cumulative and large scale process.
- **Status of issue :** Sometimes scope is decided purely on the basis of whether the issue involves statutory duties or not. Such a bland distinction may not be helpful, there are no wishful 'add-ons' in EA; whether the issue is statutorily defined or not. It does not necessarily reflect its merit in the eyes of those conducting the audit. There are many instances where a new agenda can and has been set. In this sense, EA is all about that so to begin with looking at duties it is necessary to keep in mind the responsibilities of the authority in respect of other issues.

- **Area of influence**: Strictly speaking the planned influence of the audit will be defined by the emphasis given to the different tasks in the process, e.g. RIP, PIA, MA and SoE. At the same time, all of these tasks are inter linked. In the area of recycling, for example, an internal scheme might overlap with a community-wide scheme, both could be supported by an awareness raising campaign, reducing waste and pollution would have positive effects on the quality of the environment and could lead to reduced council/LAs expenditure in these areas. While internal change might be the primary initial focus of an audit, opportunities for expanding the area of influence should be sought.

Another element in prioritizing areas of concern is to define the subject areas or environmental topics that are to be addressed. The relative importance of different categories is shown in the figure also.

Although there is some difference in emphasis between the auditing tasks, the data are even more remarkable from the degree of overlap. This supports the assertion made earlier that EA is more of a combined process than is commonly thought, more distinguishable by emphasis than by any

SAMPLES OF GUIDING STATEMENTS OF PRINCIPLES AND AIMS

Principles

- Think Globally, Act Locally
  This is probably the most commonly quoted principle of EA.
- The company/local Authority/LSG should be committed to the protection and enhancement of the local and global environment. To achieve this, policies, institutions and day to day practices will have to change accordingly.
- This authority will seek to promote the conservation and sustainable use of natural resources and to minimise environmental pollution in all of its own activities, and through its influence over others. The authority will review all of its policies, programmes and services and undertakes to act wherever necessary to meet the standards set out in the charter.

- Friends of the Earth, Declaration of Commitment
- A highly influential statement of principles (and some broad aims) which has either been adopted directly or in an amended form by many authorities.
- The LAs/company/SOE through its actions and enabling powers needs to seek to achieve sustainable
- development of resources in order to help prevent ozone depletion, tropical deforestation (and) global warming and as Environmental Strategy. The imperative of sustainable development rarely achieves such explicit recognition but it is in fact, integral to most audits.

Aims

In practice aims and principles are quite difficult to untangle. Aims should however, be slightly more focused than principles and should point in a general direction rather than represent a set of beliefs

- Assembling data about the state of health of the local environment and the authority impact upon that state of health on regular basis. Using the data to increase understanding and awareness to fulfil the public's "right to know", and to generate action to sustain and improve the environment.

These were the main aims identified by UK's local authority association survey in 1990. The first statement clearly relates SoE and the second to IA. By and large they are still applicable today.

ENVIRONMENTAL CATEGORY

There is a core of a dozen or so issues which receive much attention with waste and recycling energy transport and purchasing stinging out particularly with regard to RIP issues.

It is already highlighted preference that has been demonstrated for short-term small scale objectives this is reflected here too with the least popular IA issues being those associated with longer term change. This includes investment policy, economy and work and consumer advice and protection. These are the key areas of interest and it must surely receive more attention in the future stages in an environment auditing process. Figure illustrates the status of financial issues in the definition of scope and explores the broader resource implications of EA

Indicators and targets

As the degree of rigour demand from EA has increased over time, the issue of assessment criteria has gained prominence While an audit undertaken previous might have got away with a relatively descriptive evaluation of policies practice and environmental quality. Now it would need to include a more objective analysis in terms of how the authority performs against a range of measurable targets and standards. In a sense the balance is swinging away from an inductive and more towards a normative process. At the moment, however, authorities are left in a kind of 'wilderness'. While the demands of EA are increasing, particularly with respect to Local Agenda 21, the guidelines on targets and indicators are still being developed.

Thorny economic issues

Two of the least popular issues that are covered in EA relate directly to the economic policies and practices of a local authority. Given the widespread endorsement of sustainable development which is substantially an economic concern this lack of attention is particularly worrying The issue of 'investment' is illustrative of a number of underlying economic factors, which define scope. It is potentially a major tool for achieving environmental goals - through fund management, capital building programmes, provision of loans and grants, etc.... Reasons why investment has been largely ignored range from claims that to include programmes provision of loans and grants etc. Reasons why investment has been largely ignored range from claims that to include environment factors in financial decisions would be precluded on legal grounds (not altogether true) to difficulties in finding suitable companies and funds to invest in (hardly an insurmountable problem). A more realistic explanation probably has to be with the underlying economic interests, dominant in local authorities in a number of cases that were examined the section of the authority responsible for economic development viewed the audit with suspicion and where possible avoided participation in the process. This stance was tolerated because of the powerful backing for economic interests within the organization. This simply highlights the point that environment and economy are still regarded as strange bedfellows and wherever conflict occurs it is inevitably the economic interest that wins out. While the issue of investment is peculiar in that it introduces economic factors at an early stage in the EA process, the paramount importance of these factors often surfaces at the stage when resources must be allocates in order to pursue recommended actions. As the sustainable development debate gather force particularly with the production of local Agenda 21s, there is hope that the weight assigned to environmental and social, as opposed to purely economic, factors with increase in local authority decision making. This will inevitably involve a challenge to many traditional assumptions and powerful interests in local government.

An ideal sequence of activities in relation to a particular environmental subject area (e.g. transport or energy) the authority should set a range of carrying capacities that reflect the limit of justifiable human impact on the environment. With these capacities as the template, objectives can be defined in terms of targets, which are precise statements of what is to be achieved, and within what time scale. Progress can then be monitored using different types of performance. Naturally the sequence never runs quite so smoothly. There is as yet no universally agreed carrying capacities and it is doubtful if they're ever will be given the scientific uncertainties and political exigencies as discussed earlier. This makes defining targets an often-impossible task, which in turn leads to indicators becoming redundant. The solution as one might expect, is to concentrate on what one knows and on what one can do. For example the precise limits of water pollution that can be tolerated might bot be known but there are minimum standards defined by legislation and the damage that is resulting from what is known, should indicate the required action. In effect of this, putting into practice, the principle is that until sustainable development can be more clearly defined, action should be directed towards reducing unsustainable development.

It is just this type of pragmatic approach that is being adopted by some authorities. Figure below shows some of the sources identified in the survey that have been used to help define targets and choose appropriate indicators.

Clearly, standards derived from policy and legislation are the most accessible and useful, especially in terms of ensuring compliance. It is also encouraging to see the number of authorities that are seeking to go beyond minimum standards, towards identifying best practice elsewhere. This could be of particular relevance in areas where no legislative or policy guidelines yet exist. Thus while there are agreed standards for some aspects of environmental quality this is not the case for many areas of policy and internal practices. The checklists in this regard, are a useful set of targets and indicators for both SoE and IA. They could form the starting point for an authority to develop their own assessment criteria.

The dissemination of best practice knowledge is dependent on an effective information network. The sources and form of information in EA are discussed in the next section.

COLLECT INFORMATION AND ESTABLISH DATABASE

Information is the essential raw material of an audit. The quality of that comes with a great determination of the level of assessment and ultimately the standard of decision-making that results.

The key characteristics for the information are **relevance** and **reliability.** Both will be facilitated by the definition of operational objectives. An early scooping of the issues allows effort in collecting information to be clearly focused, while the establishment of targets and indicators points the way towards the type of data required. The alternative, of launching in with ill-defined objectives, risks the problem of information overload where the direction is lost under the weight of too much data.

'Reliability' simply means ensuring that the information is up-to-date and accurate. Efforts should be made to substantiate findings and to ensure that all the useful sources are investigated and a variety of collection methods employed. The methods, being inextricably linked with the assessment procedures, are discussed in the next section. Potential information sources may be looked at in four categories:

1. **Data already existing within the authority:** These are the most accessible, locally relevant and perhaps the most useful data available. The figure shows that practically all authorities have made use of them. Indeed the missing 10 percent must be a statistical anomaly because all auditors, even consultants, draw heavily on internal data. The figures were equally high for SoE as for IA, once again suggesting that there is significant overlap between the tasks and that there is obviously a deep well of information that can be drawn on internally.
2. **Agency sources:** Figure (information sources) also illustrates a whole range of other agencies, which can and should be investigated. It reveals that the potential of these agencies has not yet been fully realized but contacts and networks are certainly expanding all the time.
3. **Environmental forum:** An excellent way to harness the cooperation of both other agencies and the wider public is through an environmental forum. This will act as a two-way conduit for information.
4. **Documentation sources:** There is a whole host of literature now available to help in developing the EA process. Some of the earlier reference documents are shown in Figure (information sources-reference documents) the bibliography will update this list. The number of authorities who have used the 'charter' is extremely revealing of the influence this has had. The process of information exchange that clearly goes on directly between local authorities and more indirectly through best practice guides such as 'Environmental Practice in Local Government'.

Establishing a **database** at an early stage in the process allows the data that are collected to be stored in an accessible manner and subsequently facilities monitoring and updating. Undoubtedly the most useful but also the most expensive option is a Geographic Information System (GIS). However only 2 percent of those surveyed had an established GIS . Most opted for a manual approach, which is naturally cheaper but is possibly only delaying the inevitable requirement for a computerized database. The merits of different systems are discussed in Chapter

2. Failure to establish some sort of system has led in more than one case to contacts having to be retraced and a whole new body of base data reassembled - definitely an unenviable and superfluous task.

ASSESSING ISSUES AND PRIORITISING POTENTIAL ACTIONS

Assessment procedures represent the point of praxis where the information that has been collected must be evaluated against the objectives in order to aid decision making and ultimately facilitate the implementation of actions. Attempts so far at introducing technically objective procedures have not been entirely successful. It is not only that the tools for such an assessment are not developed but also many of the issues involved simply do not lend themselves to the analytic evaluation, which this implies. As a result, many audits have tended to be of a descriptive nature (see figure procedures for data collection and analysis), more compilations of issues than review of performance. The recommendations in this section are an attempt to make the most of the skills and resources currently available rather than trying to introduce a spurious objectivity through a supposedly scientifically objective process. It involves a four-step assessment, which is as follows:

Steps 1: Test significance of the data

This follows on directly from the need to ensure that relevant and reliable data have been collected. The key test is that the information is appropriate to help meet the set objectives and ultimately help advance change in the right direction. Checklist on a simple matrix can certainly help in this respect to ensure that information is adequate for the task in hand.

Step 2: Choose assessment methods

The degree to which targets and indicators have been defined will determine the nature of this assessment. If, for example, the authority has set a target for itself to convert all council vehicles to unleaded petrol, then the assessment can clearly see the proportion that are converted and recommend a programme for phased conversion in the future. This is perhaps an overly simplistic example, but the lack of a target would in this case probably result in a descriptive treatment of the pros and cons of unleaded versus leaded petrol and would lead to difficulties in recommending a way forward. Although many authorities had regard to predetermined indicators and standards. Figure (Assessment criteria) shows that the majority still had not.

As stated in the previous section, the methods for data collection and analysis are inextricably linked, particularly when the form of the data input predetermines the level of analysis that is possible. Figure (Procedures for Data collection) indicates the choice of methods.

The data clearly shows that the preferred approach to assessment was to describe rather than evaluate impacts and performance. As for the form of data, SoE tasks obviously involve a wider town for information but the majority of this is still from secondary sources. By implication, the majority of data input for IA comes from internal sources. There is an obvious need in EA to introduce a greater degree of rigour into the process. Earlier attempts has been made to do just this by suggesting a number of techniques which can help in objectifying the assessment. In using these it must always be borne in mind that they aid the political process but should not be used to supplant it. The subjectivity involved should be highlighted and not hidden. So while the best practical methods can be employed, the emphasis should always be on maintaining transparency and participation. In this way, debate will be generated, and it is this debate that will ultimately lead to more effective decision making.

Step 3: Decide on appropriate format

The most appropriate format is obviously dependent on a wide range of factors, but box approach to the assessment below presents a simplified approach that has proven quite popular in practice. With varying emphasis on the different elements it can be made to apply equally well to SoE and IA tasks.

This type of formats highlights the subjectivity and uncertainties involved in the assessment. If based on a reasonably comprehensive knowledge of these issues, it should aid participation and provide a clear basis for decision making.

Step 4: Prioritise options for action

Within the sequence presented in Box simplified approach to the assessment, the 'options for action' represent a quite discrete element of the assessment. Issues identified as being of greatest concern need to be further prioritized so that the choice of and basis for recommendations is made more explicit. The following three criteria should be of use:

- *Cost* - This is probably going to be the overriding factor. All the implications of proposed actions should be investigated so that the nature and scope of the resource commitment is fully known in advance. Staff time, in particular, can be a major cost component and should be included in calculations.

> ***Box :*** A simplified approach to the assessment For each set of issues (policies, practices or environmental topics) tackle the assessment in the following sequence:
>
> (1) General introduction and review
> (2) Current status of the issue with regard to the local authority
> (3) Statutory obligations: best practice elsewhere; possible standards, targets and indicators
> (4) The assessment of 2 in relation to 3
> (5) Options for action
> (6) Conclusions and recommendations

- *Effectiveness* - This refers to the potential to bring about real change. All actions should be considered for their direct impact, such as energy savings, and their indirect impacts, for example raising awareness.
- *Timespan* - the amount of time it takes before results become apparent is an extremely important political factor. A balance will need to be struck between short-term/small scale and long-term/broad scale actions. The immediate benefits are important, particularly as a demonstration of commitment, but should be viewed in the light on of the more broad-ranging auditing aims.

DECISION MAKING

This may be viewed as the final stage in the process. In fact it is no more than a midway point, because it is only in the subsequent stages that tangible progress has begun to be made. In fact, decision making should realistically be viewed as a continuum, with different levels and a series of implicit points. Most obviously, the members and officers closest to the audit at several different stages in the process will need to make decisions on ways forward. Choice of aims, objectives, indicators, etc. all refine the options open and therefore represent decision points. On another level, council staff and the wider public should have helped frame these choices and have thus also participated in decision making. Their feeling of 'ownership' for the process (and thereby their degree of involvement) will, as much as members' commitment and the level of resources, decide the ultimate effectiveness.

The issues involved in decision making, both internal and external, are essentially about institutional management. As such they will be addressed in the next chapter.

SPECIFY FORM AND FORMAT OF ACTIONS

When arrived at this stage where a set of recommendations have been agreed on it is now extremely important that the process presses forward so that these are put into effect. The two components that will characterise an action plan are form and format.

Form

Form relates to the worked-through details of what each particular action proposal involves. It is basically a further development of the assessment but this time not focusing on what the authority would like to do, but on what it can and will do, and how it proposes to go about it. There are four basic elements.

(1) ***Prioritisation*** : This should be looked at again, but unless circumstances have changed it should be the same as in the assessment.

(2) ***Budgeting***: While actions requiring resource input may have been previously costed, they now must be given a budget for implementation. It could either be a special 'green' could involve drawing on existing departmental budgets. The latter is probably easier but allocations should be 'tagged' so that expenditures can be monitored.

(3) ***Delegation***: The responsibilities with regard to each action should be specified. It may be appropriate to write these into service plans or personal development plans.

(4) ***Timeframe:*** Targets for what is to be achieved should also indicate when it is to be achieved by. A useful scale might be ongoing; immediate; within budgetary year; within specified medium-term period; long-term.

Format

The format for presenting recommendations might seem like a rather irrelevant point to comment on but there are in fact choices here that can very much help or hinder the EA process. The basic choices are to include the recommendations for action within the audit report (as required by EMAS) or to produce a free standing action plan. The former allows easy different between issues and proposed action, thus aiding understanding but allowing only brief details to be included. The latter allows for a comprehensive presentation but can delay the process and might not all for sufficient justification of why particular actions are to be pursued. Combining both approaches could maximize the advantages and would also facilitate a public consultation exercise by getting comment on the preliminary audit report before producing the final action plan.

IMPLEMENT ACTIONS

This highlights the need for a monitoring system but also for an implementation 'regime' which maximizes outcome. Both of these requirements will be discussed later. Figure (significant actions) gives an indication of the range of actions which authorities themselves have considered significant. They are grouped into three categories so that the areas of greatest progress can be seen.

There is a very interesting spread evident between the categories. It is not simply the tangible environmental improvements (recycling, purchasing, tree planting, etc.) that are judged as important. The longer-term institutional changes (either emanating from the EA process or as a result of training and awareness raising) are seen as significant benefits. It appears that a good balance is being struck between these different categories.

MONITORING

The monitoring task has begun much earlier in the process. Creating a suitable framework for monitoring should have been a consideration from at least the operational objective stage when a set of measurable indicators is chosen. This framework should have then been integrated into the other stages of the process (e.g. while specifying form of actions and choosing a database).

A monitoring framework should achieve two main elements:

(1) ***Checks and balances***: Feedback into the process in order for standards and mechanisms to be improved. It therefore involves monitoring the implementation of actions, the performance of policies and the management systems.

(2) ***Environmental quality and awareness***: Although of a different nature these two components could be regarded as the standard against which progress is measured. This aspect of monitoring basically involves the continual updating of a SoE survey.

While some recommendations from an audit are relatively easy to monitor, for example the inclusion of environmental implication sections in committee reports, others can involve an intricate set of procedures, for example the myriad of elements in a community recycling policy. The monitoring framework must therefore be custom made to suit each particular issue.

REVIEW

The review represents the meeting point between two phases of the EA process. There are two basic ways to approach a review. At first, there is a partial review, which may be most favourable by the companies local authorities, whereby the performance of the process is reported on and in particular action implementation is evaluated and progress updated. Secondly, there is a comprehensive review where the original aims and principles are reexamined to accommodate major institutional changes and new knowledge. It is, for the first one is on new legislation and policy guidance, altered financial circumstances, a revised political agenda or management reorganisation. Examples of the season's approaches are new technologies, advances in scientific understanding of environmental capacities or impacts, developments in best practice or the realisation of substantial gaps in the current process.

While the partial reviews should be reported on at least annually, a comprehensive review might only be needed every five years or so. The results of both should be widely disseminated and can play a major role in promoting environmental awareness by advertising the benefits of positive action.

CONCLUSION

This chapter has explained a phased and strategic process whereby an EA process can be given the time and resources to evolve within an authority. It has been suggested that EA should not only be owned by a local authority/company, but it should involve the wider community.

The series of stages discussed here should help maintain the coherence of a continually evolving environmental strategy. Next chapter will explore the elements in a complementary management system.

CHECKLIST

- An EA should be viewed as both an internal and external process. The tasks involved may be broadly distinguished as IA and SoE. Every audit will address both tasks, but with a varying degree of emphasis on each.
- Adopt a strategic approach to EA. Within a staged process, focus objectives and scope towards the achievement of tangible environmental improvements. At the same time maintain a corporate overview of the entire process.

- The principles and aims should be as 'strong' as is realistically possible (given resource constraints, levels of commitment expected, etc.). A vision of sustainable development could be set out and expanded upon.
- The aspirations of the audit should be tempered by specifying realistic but achievable 'general' objectives. This will help fulfil the aims while keeping the process manageable.
- Clearly define the scope of the audit at an early stage. This should specify the issues to be addressed and the depth of analysis required. Choose the most reliable and challenging standards, targets and examples of best practice. Define issues in terms of measurable indicator.
- The assessment should interact with all other stages in the EA process. Best practical methods should be used to aid objectivity but should not obscure the transparency and political issues.
- If EA is viewed as an interactive and political process, then decision making will benefit from participation and transparency. This implies addressing all levels and points at which decisions are made.
- A set of implementable action must result from the process. Spend time specifying the appropriate form and format.
- Actions should be aimed at bringing about both short-term and long-term changes. Maximize outcome and not output.
- From the outset consider the elements in a monitoring framework. Later, after action recommendations have been made, elaborate on each of these elements in order to provide a sound basis for reporting progress.
- A system of partial and comprehensive reviews should continually update the process and report publicly on progress.

CHAPTER 14

MANAGEMENT OF THE EA PROCESS

INTRODUCTION

Issues of management have been referred to previously as being at the very core of an EA process. In many respects it is management, in terms of the organisation, the individuals and the specific mechanisms, which defines an audit rather than the scope, procedures or other aspects of the 'technical' policy process. To put this another way, there is a distinction in auditing between the techniques of environmental management and the political process of decision making and institutional change. This chapter aims the discussion of process by examining how EA can or can be made to fit within the structures of a company/local authority and how it can promote participative decision making and positive change.

While obviously no two local authorities/companies are the same, they all share certain characteristics defined by the political and legislative context. So, while it is true to say that there can be no definitive way to manage EA, it is equally the case that certain elements will be common to all. These communal elements form the basis of this discussion.

THE POSSIBILITIES AND PROBLEMS OF A MODEL MANAGEMENT SYSTEM

There is always a danger if recommending approaches to management in away, that the discussions become too prescriptive. The emphasis ceases being on workable elements and instead puts forward a definitive and one dimensional 'model' of what a management system should look like. Just this type of approach is the basis for much management theory. The elements in a system are combined and sold as a 'recipe' for success which can be taken and applied anywhere. This approach to presenting a model management system is not pursued here. This chapter takes as a basic rationale that an EA process is inextricably linked to its organisational and situational environment. The degree of fit and the scope to evolve are therefore the prime determinants of a successful approach.

The model presented below Figure EA management system is not intended as a prescription for how to organise and manage an EA. It is more a description of a set of elements or mechanisms that have been found to work well in a number of situations and which should be replicable in most local authority contexts. Alongside the tangible management mechanisms it also illustrates some of the intangible elements which will be of equal, if not greater, importance in a successful EA process.

Each of the elements in the model will be elaborated on in the following section. A series of recommendations are made based on the evidence of what appears to work in a wide number of cases. Throughout the chapter the discussion will attempt to address some of the more intangible factors - such as values, perceptions, and feelings of ownership - which to a large extent will characterize an audit.

It should be kept in mind that although management issues may be couched in terms such as mechanisms, duties, procedures, etc. they are in fact about people. Ultimately it is the individuals involved who will determine the effectiveness of the process.

CORPORATE COMMITMENT AND COMMITTEE STRUCTURE

It is at the member level that the decision to undertake an audit is ultimately taken, so the support and commitment of the members will obviously be crucial if the process is to have any chance of being, successful. Just to illustrate this statement, Figure (areas of failure) shows those aspects, which the questionnaire respondents thought most impeded progress with their audit.

Lack of direct member commitment and of the resources that would be one sign of evidence of this are cited as the two most common causes of failure in EA.

The most visible manifestation of this commitment is in the committee, which deals with the EA. Figure (dealing with environmental initiatives) chosen.

The LA/company may establish special environment or 'green' initiatives committees specifically to deal with the audit and other such corporate strategies. These committees may have a pivotal role in EA. They may act as a forum for debate, an arena for progress to be monitored, a vital link to other company/LAs policy areas and to the corporate goals, a political expression of commitment and a mechanism to ensure objective assessment and effective implementation. They also help ensure that the audit is managed as a cross-authority initiative.

In absence of environment committee the audit must be added to the remit of another committee. The danger with this is that the audit could receive low priority, being simply one additional matter added to their workload, or it could be marginalised within the authority by being viewed as the responsibility of one particular department.

The other level to policy direction is to inculcate the principles of the audit within all sections of the authority. This can be greatly facilitated by requiring all committees to have regard to the environment in decisions that they make. This will in any case be necessary because it is through the committees that recommendations must be passed for approval and resource allocations. The more involved these committees are, possibly through representatives on the environment committee, the more likely they will be supportive of the changes the audit seeks to bring about. A relatively simple mechanism, which many authorities have found useful, is to include a paragraph on the environmental implications of each decision within the committee reports. This is already a requirement with regard to financial and equal opportunity concerns and if undertaken in the correct spirit can be extremely effective. Another way to reinforce a corporate commitment is to include the broad aims of the audit within any statements of overall policy.

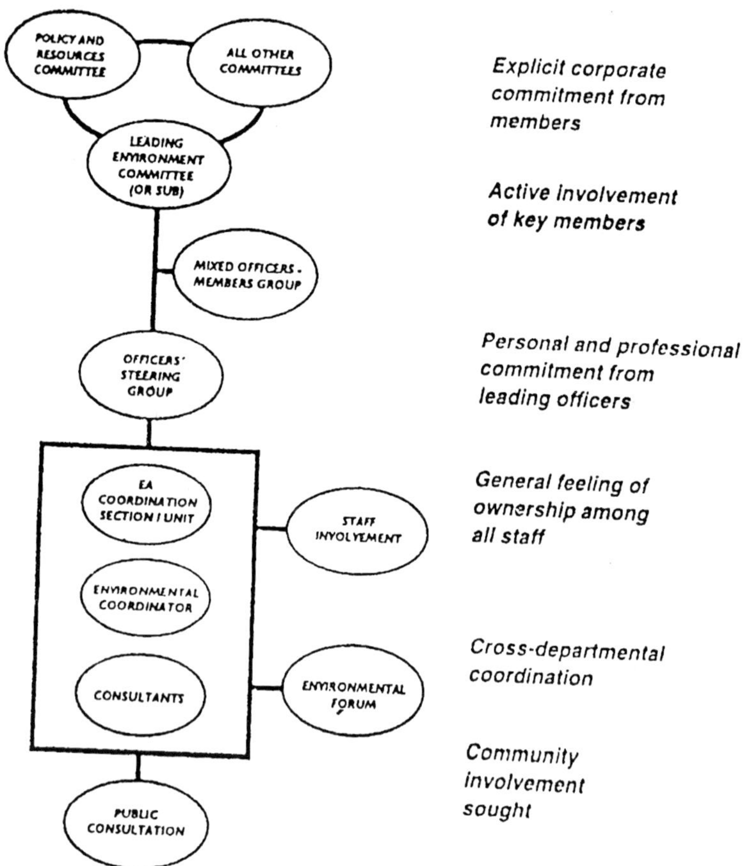

Fig. 14.1 : Elements in an EA management system

OFFICERS' STEERING GROUP

The LA/Company authorities may create a steering (or working) group as the principal interdepartmental management mechanism. This will certainly provides a clear control structure. In fact, much more than that. It acts as a forum for debate, direction and dissemination of information. More specifically, it performs the following functions:

- it enlists the participation of all departments and other interested parties in the management of the audit;

- in the absence of another coordinating mechanism, it provides a formal link between officers and members;
- it advises and manages those undertaking the day-to-day work on the audit;
- through its members, information and comment can be passed both to and from all levels and sections of the authority
- important issues can be debated and reported on in a formal way;
- it oversees the work on all stages of the audit process;
- it monitors progress with implementing actions and facilitates updating and review,

The exact terms of reference and composition of a steering group vary widely. What is important is that the membership is drawn from throughout the authority. Preferably they should be high-ranking officers but more importantly they should approach the audit with a personal and professional commitment. Authorities/ companies have benefit of having a more open steering group's structure. Direct involvement by councillors obviously helps improve communication, while representatives drawn from local community and environmental organisations strengthen the public's participation. Occasionally a second working group may be created comprising more junior officers, but also other staff and public representatives. This can relieve the main group of some of its duties while also building up a broader base of auditing expertise and generating wider involvement and ownership.

Finally, the steering group must be a permanent addition to management structures. This means staying in place even after the initial audit phase. During certain periods, it may not need to meet as frequently but it has an ongoing role, which needs to be performed.

MANAGING THE PROCESS - INTERNAL AND EXTERNAL EXPERTISE

It has already been discussed earlier that who should do the audit? While there are no definitive answers, a combined Approach using internal and external expertise would seem to be the most appropriate option. This section will examine the mechanisms

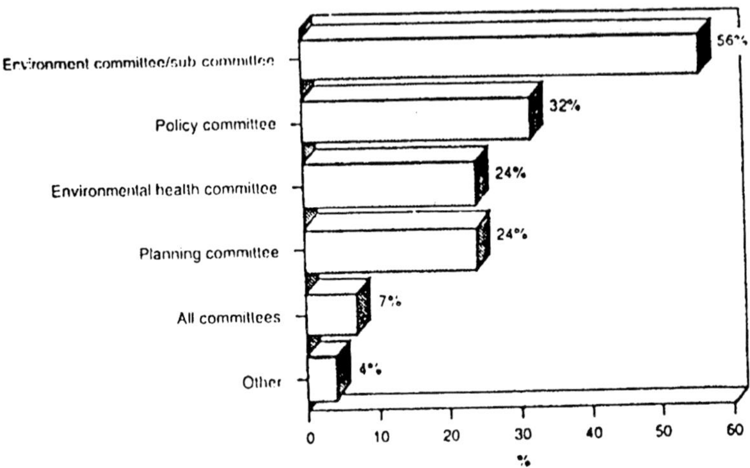

Fig. 14.2 : Members committee dealing with environmental initiatives

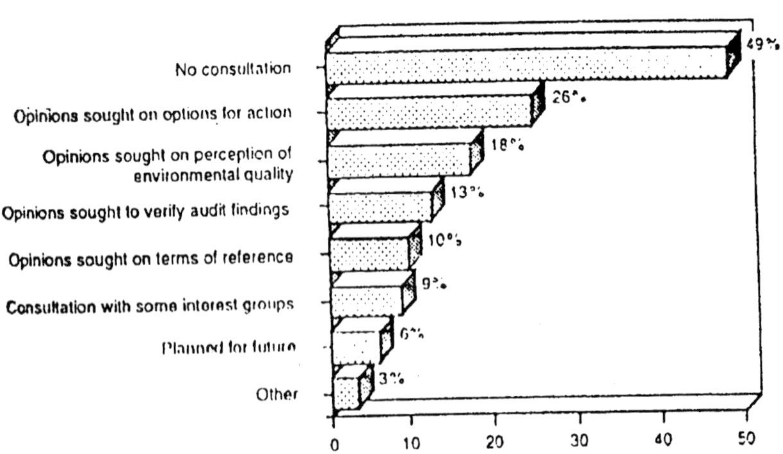

Fig. 14.3 : Public consultation

for day-to-day management in more detail. Looking at the general in-house arrangements, the specific role of an environmental coordinator and the benefits to be gained from using consultants.

In-house arrangements

The benefits to be gained from conducting the audit predominantly in-house are very convincing. The central point is that a process must be managed by those involved in its formulation and most importantly its implementation. While theoretically it may be possible for a committee to formulate the aims, hire consultants to conduct an assessment and pass the results on to staff for implementation, this is not an EA process and for this reason will almost certainly be limited in its usefulness. The process is as much about how it is done as about what the outcome is. The change an audit as in the wider environment. This means developing robust internal mechanisms form an early stage. Some of the advantages of this will be:

- As an ongoing process requires that expertise and awareness are built up over time. The sooner the learning process starts the better.
- Widespread involvement is essential, in the first place to overcome any resistance to what might be seen as outside prying and in the second place to being generating a feeling of 'ownership' among all staff and hopefully the public.
- By examining their own activities individuals can learn the importance of their role and responsibilities in relation to the environment. This awareness is an essential prerequisite to achieving a change in values.
- More pragmatically, an in-house audit is probably the cheapest option and certainly the most flexible.

Naturally there are also disadvantages in this approach, but as it is argued below, these can be compensated for by the limited employment of consultants.

Having decided to manage the audit in-house the next questions to arise are where and how.

Who should be involved in the in-house audit?

The greatest difficulty that may be arisen is that an EA by its nature is cross departmental. It should therefore ideally not

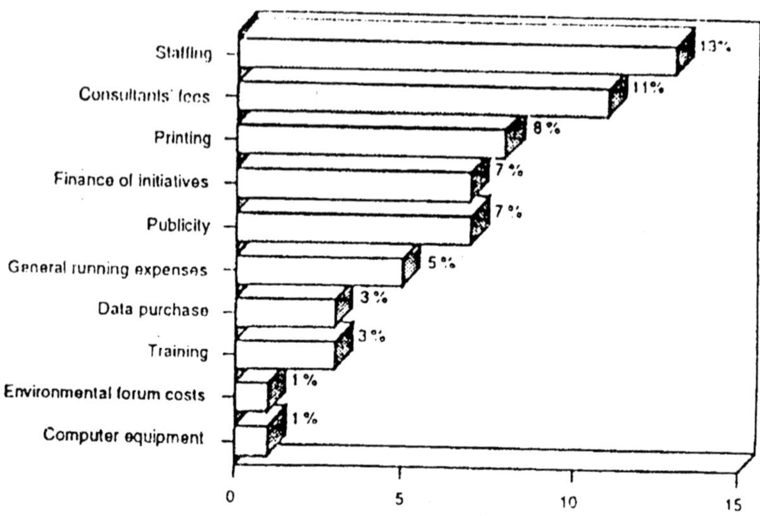

Fig. 14.4 : Budgetary elements

be done by one section but should be incorporated into the work of all departments and functional areas. Naturally this requirement is difficult to meet. Being a new process, EA must, at least to begin with, fit into the existing administrative structures. It would be argued that this is the only viable approach to avoid the audit becoming marginalised as a process unconnected from the rest of the work of the authority. Figure (nature and role of the EA unit) shows the responses of those authorities that said the audit had been undertaken by a special EA 'unit'.

Obviously what is taken to constitute a unit varies considerably but it does indicate the areas of the authority where the management has concentrated. Of particular note is the split between the planning, environmental health and chief executive's functional areas. This same division is evident in figure (dealing with environment initiative) which showed the committees dealing with EA (note: chief executive's office is considered to be responsible to the policy committee). While the chief executive's office operates on a corporate basis, the planning and environmental health functions are quite distinct and often professionally guarded interests. This can sometimes give rise to unhelpful tensions. In a particular authority that was surveyed, two totally contradictory responses were received, one from environmental health and the other from planning.

In the long run no doubt EA will find its place in the organisation. The environment committee and the steering group is already established elements in an environmental management system. In addition to these, EMAS proposes a whole range of mechanisms for conducting and reporting on audit performance. Some authorities may have already instigated major organisational changes in order to give greater prominence to environmental

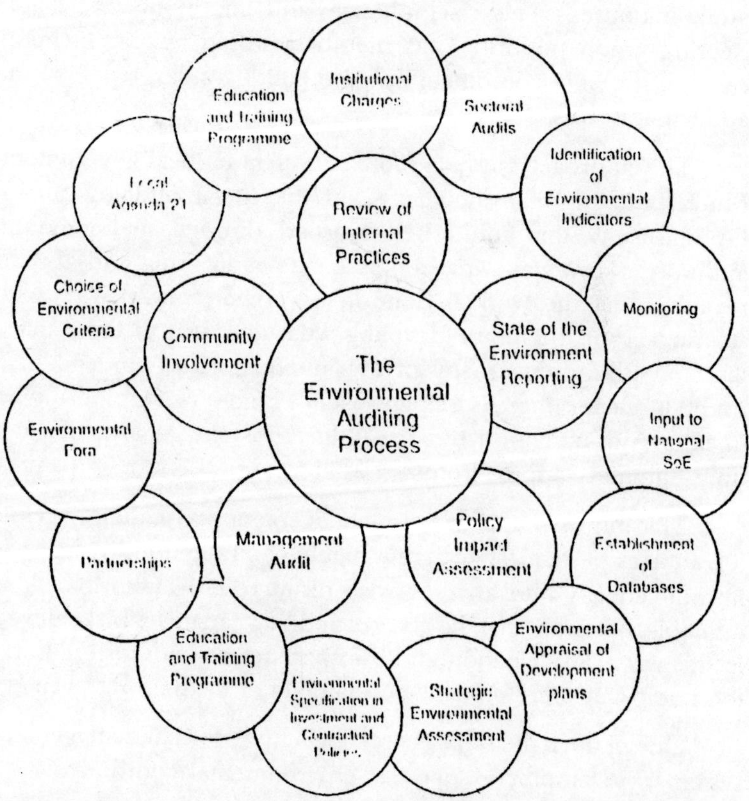

Fig. 14.5 : Environmental auditing at the centre of an overlapping system of environmental management

matters. Many of the environmental committees and environmental service departments may not just involved with EA but can be the result of an amalgamation at both officer and councillor level of separated environmental functions. Typically this will include planning and environmental health, so the departmental tension may well resolve itself through a broadening of the remit of specialist sections and a growing realisation of the interdisciplinary nature of environmental problems.

Managing the in-house audit

Figure (Role of the E/A Unit) above gives a rough indication of the type of arrangements that have been used in managing an audit in-house. This re-emphasizes the important role that a steering group performs. Its members would appear in many cases to not only coordinate but also conduct a large part of the audit themselves.

This approach is much more prevalent in local government. Much time in total should be on the initial auditing phase. Optimistically this could be absorbed through the company authority. However, what the case studies in India and abroad reveal is that the work is done in a very piecemeal way, with steering group members devoting what amounts to their space time on performing many of the auditing tasks. Indeed in a number of cases there is a single, usually high-ranking officer, who in effect ends up conducting the audit almost single-handedly with only minimal support from elsewhere.

This informal approach to management may certainly get its advantages particularly where a highly committed individual has introduced EA into an otherwise disinterested authority. It is arguable, however, that the goals of auditing can never be achieved in this way. Participation and transparency are not fostered and the audit can therefore not be maintained in the long term.

One of the main steps in creating a more formal management system is to employ a specialist environmental coordinator.

The role of the environmental coordinator

The figure has already shown the importance of having a specific environmental coordinator in order to conduct the EA tasks. Earlier it is discussed the typical job description for a coordinator post. Clearly the range of duties should more than justify the expense involved, and with Local Agenda 21 and FMAS soon to be added to these duties it is arguably an essential investment.

Particular care must be taken with regard to the location and rank of the coordinator's post. It should obviously be situated so that interdepartmental involvement is maximised and the greatest numbers of links are forged. With regard to the previous discussion, this would most likely be an amalgamated environmental services department or otherwise the chief executive's office as a relatively neutral point. With regards to rank, the post should be sufficiently senior so that the wide range of duties can be carried out with authority and without having to constantly seek authorisation.

The emerging profession of environmental coordinators is qualitatively different from most other types of local authority professional groupings. They essentially perform a linking role between a whole range of natural and social scientist specialists. Their usefulness is therefore not in doing all the work on the audit themselves but in enabling others in the authority to effectively manage the process.

POSSIBLE DUTIES AND RESPONSIBILITIES OF AN ENVIRONMENTAL COORDINATOR

(1) To coordinate the monitoring, development and implementation of the City Green Charter.

(2) To develop and maintain a specialist knowledge of environmental policy

(3) To coordinate the development of a Local Agenda 21

(4) To coordinate the development and implementation of an Eco-Management system

(5) Prepare Green Charter Action Plans and monitor their implementation.

(6) Prepare committee and management reports as appropriate

(7) Contribute generally to the operation of the development team
(8) Maintain records of action taken and conduct correspondence in connection with the duties undertaken.
(9) Provide such support and advice as from time to time the health and green initiatives manager may require.
(10) Attend such committee or other meetings as may be required from time to time and give reports and Presentations as appropriate.
(11) Liaise with other directorates and external organisations as appropriate
(12) Observe the City Council's Equal Opportunities and all corporate employment policies and the application of their principles in all aspects of the service provided by the postholder.
(13) Be responsible for own safety and not endangering that of colleagues/others in workplace.
(14) Ensure that output and quality of work is of the highest possible standard and accords with current legislation and City Council policies
(15) Carry out such other duties as may reasonably be required in relation to a post of this nature, without prejudice to any grading appeal rights.

Authorities may employ a single environmental coordinator and later expand into a small team to cater for the ever-expanding EA related functions.

How to use consultants?

The two major potential pitfalls of in-house auditing are a lack of objectivity and of expertise. While these can be lessened over time by developing the internal management systems there will probably always be a role for consultants in these areas. In relation to objectivity, external auditors can bring a professional detachment and impartial attitude with them into the process. Indeed, one of the great advantages of EMAS is that it will require this external validation, thereby instilling more confidence in the level of assessment and the results. But the consultant's role should be limited in this respect to a review of what is otherwise an internal exercise. A deeper involvement in the management would not increase the level of objectivity but would certainly reduce the participation of company/LAs and thereby lessen the 'ownership' felt for the audit.

Expertise is essentially what the authority requires when it gives assignment to a consultant. This will hopefully be of both substantive and procedural issues. If consultants are used judiciously, they can help supplement internal expertise and at the same time train local authority staff in the methods and procedures of auditing. This might for instance involve an initial training course for members and officers on defining the aims and tasks of EA appropriate to that authority. More broadly there could be awareness raising sessions at all levels, from subject specific best practice to the role of individuals as members of the community. This type of training could be a continuous part of staff development.

GENERATING CORPORATE OWNERSHIP

One of the goals of EA should be to create a strong sense of ownership for the process throughout the authority. Ownership is about values, perceptions and motivations, none of which are factors that lend themselves to easy analysis. However, it is such an important issue that some attempt must be made to find out how it can be generated and maintained. However, it is such an important issue that some attempt must be made to find out how it can be generated and maintained. This section will discuss the two factors, involvement and awareness, which are elements of ownership but which represent relatively tangible aspects which should be more amenable to influence by the auditors.

Involvement

Involvement is essentially a function of participation and manageability. While the participation of everyone in the process might be the ideal, this would be practically impossible because of problems of management, consultation and the need to maintain a degree of control. The previous sections have dealt with the mechanisms for involvement of members and some high ranking officers. While each of the groups performs an essential role, they obviously only represent a tiny proportion of the local authority numbers. The kinds of mechanisms used to gain wider staff involvement are shown fig (mechanism to gain involvement of member).

THE VALUE OF CONSULTANTS

Authority X employed a consultant to carry out a joint SoE/IA on the basis of his widespread knowledge of the environment and the auditing process. The work was to take six months and lead to the production of a comprehensive report including recommendations. The audit actually ended up taking twice as long and the final report was criticised for not saying anything that would not have happened anyway. ***Authority Y hired a group of consultants to do their IA. Because of the authority's lack of knowledge of auditing and the consultants' inexperience there were problems defining the terms of reference and the appropriate fee. In the end, the audit cost more and took longer than either the authority or consultants had expected. There was also a feeling that council employees ended up doing much of the work themselves because of all the information they had to collect to give to the consultants.*** Finally,

Authority Z chose a makeshift consultancy, which could draw on a wide range of expertise to carry out an IA. Different expectations of the process led to complications, which resulted in little use being made of the final report. One of the conclusions to be drawn is that consultancies are on the same learning curve as local authorities. As experience in both the authorities and the consultancies increases there is less likelihood that these sorts of problems will occur. There are moves currently underway to create a professional accreditation body, which should guarantee standards of performance and make the choice of consultancy firms much easier for local authorities.

Most of these are fairly traditional, but still potentially effective and easy to use. What is meant by 'informal' involvement is not exactly clear. It would appear that this covers such things as word of mouth dissemination of information, a general openness to staff suggestions, relying on managers to inform staff and in turn represent their views, and more broadly, persuading staff to examine their roles and responsibilities.

Many authorities have obviously seen the benefits of having departmental 'green' representatives. Occasionally these are the same people as those on the steering group but in order to spread the workload it could be a good to involve other officers, selected or elected, because of their high level of awareness and commitment to environmental issues. They could, for example, be sub-group of the main steering group who deals specifically with maternal interest.

Finally, direct involvement of staff through meetings, presentations and training programmes probably offers the greatest, but as yet under utilised potential. The examples of two useful approaches to training are earlier discussed.

A simple method to begin is to include an environmental element in staff induction courses, although obviously other methods are needed for existing staff.

These two examples of training programmes show that on one level they are aimed at building expertise but they have a much more important role in engendering awareness. Ultimately this is how values are changed and is therefore the route to achieving the broader of EA. As Figure (Areas of success) shows, this point is not lost on those authorities engaged in EA, with raised awareness considered to be the area where most progress has been made to date.

Individuals must be shown the means and the rationale for taking action. All levels of staff will be responsible, in their different ways, for implementing the recommendations of an audit, so efforts must be made to gain their involvement and support. Figure (change of environmental awareness) gives an idea of the level of success achieved so far in raising internal awareness.

Examples of training programmes It is believed that getting everyone involved, or gaining wide ownership, is a key factor in the success of his or her audit. At an early stage a training course was organised for key officers and members who helped them in developing the audit methodology. Later, after the process had begun, a wider programme of staff training was initiated. The goal was to create internal awareness by stressing the importance of the environment. The method was to train a small number of staff, initially in two departments, and for these staff then to subsequently run training sessions with the materials provided. In this way, awareness and knowledge should 'cascade' throughout the whole authority. Two levels of training and awareness may be organised. In order to increase members knowledge and commitment, a limited number of sessions on specific environmental topics may be introduced. A larger training programme may also be run with the intention of training a large body of staff in the principles and techniques of EA and the range of participation. A large programme of awareness raising sessions may also be planned.

The final point on owner ship and involvement refers back to the point made at the beginning of this section about the need to keep the process manageable. While it is unlikely to fall down due to over-involvement, the elements in the process must still be coordinated so that effort is not wasted and tasks duplicated. The management of change requires strong support structures. Most of these structures have already been mentioned-an environment committee, steering group, environmental coordinator (or team) green representatives.

COMMUNITY INVOLVEMENT: EA FOR THE PEOPLE AND BY THE PEOPLE

Although it might sound to some authorities of LA/Company. Every audit is done for the wider community. Even those authorities that see no role for the public in 'his or her' RIP are mistaken in thinking that it is not done for the community. There are at least four good reasons why this is desired :

(1) A local authority/company represents the view of its community. On one level this can be achieved though elected councillors but more fundamentally there should be direct means of expressing views on particular environmental issues.
(2) A local authority or company should lead by example. The public must therefore be kept informed of what the company is doing and why.
(3) A local authority's / company's mandate for action ultimately comes from the public. The publics support must be sought and maintained.
(4) The ultimate goal of EA is to improve the environment. This involves raising public awareness, harnessing the community's energy and directing positive action.

These reasons should be enough to convince any authority of the need to gain the public's involvement. But, whatever the case, during EA as a voluntary activity with no mandatory requirement to involve the public are limited. Elements PLA have already been incorporated within new regulations to appraise development plan policies, which of course includes a large element of consultation. Strategic Environmental Assessment (SEA), which is shortly to be the subject of regulation, will require a new openness on the part of government agencies to inform, and involve the public. The public have already gained access to a wide range of information held by public bodies as result of the 'Freedom of Access to Environmental Information' regulation. EMAS contains requirements for public involvement. Under the scheme the published 'Environmental Statement' is intended to be comprehensible and accessible to the public. It is not compulsory but is considered appropriate to produce a draft of this first for public consultation. Most fundamentally, the principles of EA now also enshrined in Agenda 21 are based on the need for openness, honesty and participation. The concluding chapter will discuss, what this might mean for EA in more detail but one thing is certain, the community's role in setting the agenda and managing the initiative is set to grow.

The different roles of the community were briefly mentioned previous Chapters and discussed in more detail. These were basically as sources of information and advice and as implementers of change. Here we elaborate on these roles in the context of the management functions of the authority/company.

The public as informants and advisers

Referring back to the previous chapter, it was shown that local community and environmental organizations were a major source of information. Depending on the issues, there are a whole range of groups that should have useful information and advice to offer. In addition, the contacts that are made should help ensure that the audit addresses those issues considered to be of greatest importance and thereby helps defray potential criticism later in the process. Consultation with the wide public has also taken place, as shown in Figure "Public Consultation'

This has not been as widespread as one would hope. The greatest area of involvement has been post-audit, when public opinion was sought on options for action and to verify findings. Authority may be considered as a peripheral issue could for a large proportion of the public be of immediate importance. Not alone does the authority risk criticism for not addressing these issues, but it could alienate large sections of the community.

The public as implementers of change

An equally important role for the community is as partners with the LA/ Company in implementing actions and monitoring progress. This role reinforced by the first, because if the public do not feel that their involvement has been sought they will in turn be very reluctant to support the audit. This is once again an issue of 'ownership'. Apart from a limited RIP, most of the potential actions that can result form an audit will rely on the community for implementation. Authorities can make policy decisions to educate, enforce, encourage, etc. but their own actions will of necessity be limited. In a working partnership with the community the audit will become a much more powerful force for change. The sort of networks, this involves are already in place and achieving great success. A few general examples are shown in succeeding pages.

PUBLIC OPINION SURVEY

THE EACH OF THE ENVIRONMENTAL ISSUES LISTED BELOW. PLEASE INDICATE WHETHER THE ISSUE IS IN YOUR VIEW: VERY IMPORTANT: IMPORTANT: NOT IMPORTANT.

(Please tick one box for each line)

Please also add in the left-hand column any other issues that should be given priority and also star in the right-hand column those three issues which you think should have top priority in any environmental action plan in the country

	Very Important	*Important*	*Not Important*
• Energy conservation	—	—	—
• Env. Education	—	—	—
• Green belt	—	—	—
• History Building Conservation	—	—	—
• Nature Conservation	—	—	—
• Open Space Projection	—	—	—
• Pollution Control			
• Recycling Initiatives			
• Transport Policies			
• Waste Management	—	—	—

Note : This pre-audit survey was only one part of a wider programme of consultation that involved public meetings, exhibition, workshops, comments on draft audit report and draft action plan and an environmental forms

Another area where the community should be directly involved is in monitoring and reviewing progress. Ultimately it is the public and not any objective measures of performance which will decide whether the goals of EA are being achieved. Two important aspects of this are environmental quality and levels of environmental awareness. Both are directly related to people's everyday lives and their cooperation in deciding whether change has occurred or not is essential. Their views and perceptions should of course be balanced against more intangible and perhaps long-range

PARTNERSHIPS BETWEEN LOCAL AUTHORITY AND THE COMMUNITY

- Wildlife groups throughout the country undertake a range of nature conservation tasks on behalf of and with the support of local authorities. This includes nature reserve management, ecological surveys, environmental monitoring, species protection and habitat creation.
- Colleges and universities link up with authorities to provide education and training for staff, specialist advice and placement posts
- Recycling schemes run by voluntary groups often operate in cooperation with council waste disposal Services and receive financial and logistical support
- Private enterprises from industrial firms to farms, often enter into voluntary agreements to help protect and enhance the environment. This can range from negotiated arrangements with developers to improve landscaping or reduce pollution to management agreements on farms to preserve a special site of interest
- Civic societies and residential groups often advise on and help finance the enhancement of urban areas

factors but they should nevertheless be a major element in reviewing progress. More directly, specialist interest groups are often the most reliable and comprehensive sources of information on environmental trends. In addition, academic institutions, business organisations and public agencies - all have data and expertise that can help monitor environmental quality and reflect progress in order areas.

Finally a sustainable community must be built on self-awareness and empowerment. Figure (change in environmental awareness in LAs region) shows that some success can be achieved through EA in increasing environmental awareness among the public.

There is still much work to be done in translating awareness into positive action but, through community involvement, a local authority can help achieve change not just for the people but by the people.

The role of an environmental forum

Of the range of mechanisms that can be used to encourage community involvement perhaps the most significant is the environmental forum. Essentially it will a mechanism for liaison between the local authority and its community and between different interests within the community. As such it is an excellent vehicle for achieving the information, advice and partnership roles outlined above. In some regions it has been the first opportunity that environmental interest groups have had to engage with each other, let alone with the authority. The potential is therefore immense. Networks of information and contacts can be established, new alliances forged and joint initiatives developed. Most importantly it provides an opportunity for debating issues and deciding agendas. This discursive and consensual approach must surely help dispel prejudices by creating greater understanding of the limitations as well as the opportunities that there are for action. This could be particularly helpful for the local authority whose limitations on action, as dictated by mandate and finance, are not always appreciated by outside bodies. The forum should not become a venue for the public to constantly criticise the authority on particular issues, nor one for the authority to simply explain away their inactivity or lack of success. The important characteristics to foster are open and free-ranging debate. There are plenty of models of successful environmental forums.

A final point to make is that the authority should put great effort into persuading members of the public to get involved. An apparent lack of interest on the part of the community should not be taken at face value. Those authorities who have done well in EA have often had to spend time encouraging the public that their input will be truly valuable. Signs of inertia should be combated rather than blindly accepted. Thus begins the process of empowerment.

RESOURCES - MYTHS AND REALITIES

It is well said that the major barrier to the adoption of comprehensive audit procedures in local government is one of cost.

TIPS FOR A SUCCESSFUL ENVIRONMENTAL FORUM

- Membership should be drawn from as wide a range of interests as possible. This should include local Conservation, environmental and amenity groups; the private sector; educational establishments; government Agencies; other local authorities; individual members of the public with special interests; minority groups; Relevant sections of the authority itself
- If there is a high level of involvement, consider setting up sub-groups to focus on particular topics - For example: economy, wild-life, energy, etc. It is still essential to maintain the overview of the main forum
- The council's role should not be one of leader but rather of facilitator. The agenda should be set by the participants and if there is a chairperson it should be an elected post, or perhaps held on a rotating basis. The council should be careful not to over-represent itself. The facilitator role could include providing the meeting place, the secretariat and the promotion
- The frequency of meetings should be decided jointly but should be often enough for participants to feel a real Input and not so often to deter attendance. Flexibility is also important so that key stages of the audit and Important issues can be dealt with as they arise

This is undoubtedly a realistic assessment. Fear of resource implications is mentioned in the majority of texts on auditing as the principal reason for not carrying out an audit. (It is also not surprising, given the costs of some of the most well known audits - Lancashire County Council in UK, for example spent 200,000 pounds on their SoE survey while Humberside County Council's (UK) pollution audit cost 300,000 pounds).

However this is essentially an issue about commitment. Resources are in fact, tangible evidence of a real will that act on the part of those directing the audit. The myth about resources in EA is that it is an expensive process to manage. As explained below, the reality is that the process is relatively inexpensive to undertake and manage but will inevitably highlight the need for extra expenditure on environmental objectives.

Initial budget shows the initial audit budgets for surveyed authorities and figure the implementation budget. The major budgetary elements are shown in Figure Budgetary elements.

Initial budget

The initial budget represents the costs involved in getting the EA process off the ground i.e. the surveying and reporting costs. The audit may be financed from existing resources or/and of specified budget. This redirection of existing resources towards the audit raises questions as to where the money would have been spent if not on the audit. However, in effect the allocation of a special EA budget may involves a redirection of funds within an authority. This is the reality of auditing. As with many other areas of environmental activity, such as recycling, tree planting and nature protection, the authority must decide if, on balance, the costs are justified. Where it differs from other areas of environmental activity is that the systems set up by an EA management system should help an authority coordinate its environmental expenditure. Thus, for example, the rationalisation of recycling activities or the pursuit of multiple objectives through single activities can improve overall cost-effectiveness. For this reason clear knowledge of the costs is essential. Without this knowledge it will be very difficult to evaluate the cost effectiveness of the auditing process and to justify its continuance.

The amounts in Figure 15.10 clearly show that auditing need not be as resource consuming as is commonly through. The majority of authorities spent between pounds 10,000 and pounds 15,000 on the initial audit. Figure shows that by far the largest cost element is labor, staff and/or consultant's time. The other significant cost elements are printing, publicity and general operation expenses. These costs may be identified and clearly managed within the EA process. Given clear objectives for the audit and knowledge of available resources, the process can be made to fit within these parameters.

Implementation budget

The actions recommended by an audit will not involve cost implications. These may have to do with changes in policy and stall priorities or simply rationalisation of activities. Other actions may incur minor costs but these are often only in the short term and are more than offset in the longer term. Recommended actions on energy efficiency, transport management and recycling are examples of areas where long-term cost savings are possible.

Implementation budget represents very specific 'green' projects such as a new tree-planting scheme, purchase of additional monitoring equipment, or establishing internal mechanisms for recycling. The larger amounts are a range of revenue and capital expenditures which will forward the objectives of EA but to a large extent represent costs which would have been incurred anyway. This usually includes community recycling, traffic calming, grants to environmental organisations, landscape improvements, litter abatement and so on. There is a distinction here between the audit as a vehicle for change and as a mechanism to coordinate existing policies and practices. More broadly, this could be perceived as a distinction between the pursuit of sustainable development and the continued improvement of existing (traditional) environmental areas. It is obviously realistic for an audit to attempt to coordinate existing environmental expenditures but it should not limit itself to this. This need not necessarily involve additional resources; more important is that policies are integrated and resources are used to pursue multiple objectives. Where resources savings are achieved, such as in those areas listed in column one of table potential actions to improve EA. The net benefit can be ploughed back into other EA activities. Some authorities may set up special funds for just this purpose.

A final point is that the non-response rate to questions on resources is very high, which indicates a possible lack of through for the financial implication of EA and an apparent weakness in monitoring the management of the process. Together with data on the amount of staff time put into EA, this is an area, which will have to receive more attention, particularly in the context of EMAS.

CONCLUSION

This chapter has discussed the elements in a management system for EA. These are illustrated in Figure public opinion

Table 14.1 : Potential actions to improve LA's practices

Savings	*No cost*	*Low cost*	*High cost*
Energy efficiency Measures	Assess environmental impact of all policies and practices	Staff and public education and training	Redesign or relocate council building
Rationalize Purchasing	Greater corporate commitment and initiatives	Publicity of environmental issues	New transport networks and modes
Community Involvement in Implementation Initiatives	Generate staff 'ownership' of environmental	More environmental monitoring Centre	Environment/ Ecology/Interpretive
Switch to unleaded or Diesel petrol	cycle allowances. More incentives to Use public transport - e.g. contributions to bus and rail passes	General environmental – e.g. tree planting, pocket parks, wildlife meadows	Major environmental improvements - eg creation of country park, wildlife areas
Fuel efficiency Measures	Greater dissemination of environmental sensitive way	Grants for voluntary organisation	Employment of environmental specialists

Table 14.1 : *Contd.*

Savings	*No cost*	*Low cost*	*High cost*
Reduced car Allowances Reduced essential Car users Reduced engine sizes Parking charges	Council land and buildings managed in an environmentally sensitive way provision	Support community initiatives education and information	Invest in GIS for monitoring,
Reuse and recycling of all appropriate materials	Include environmental specifications in contracts	Service on environmental forum	
Reduced material use -e.g. sending E-Mail rather than memos reduces energy and paper use	Inform business and the public of impacts and responsibilities	Purchase of environmentally friendly goods	

Table 14.1 : *Contd.*

Savings	*No cost*	*Low cost*	*High cost*
			Improve non-car access to council land and buildings
			Traffic calming schemes
			Support public Transport – eg subsidize buses
			Assess environmental impact of all developments

Notes :

(1) Cost is calculated on an aggregate basis. If for example, short-term (1-3 years) minor costs are offset by greater long-term savings, this is considered a saving.

(2) Calculations exclude staff time while recognizing that this may be a large cost element

(3) From left to right most of the actions could be placed in the most category with a proportionate increase in effectiveness and expenditure

survey. The tangible elements to the system are drawn from emerging best practice in a wide number of local authorities. The intangible elements cover a range of factors, which have been demonstrated to be of fundamental importance in the conduct of an EA process. An authority undertaking an audit should assess the contribution that each element could make to their process but should only choose those that are appropriate to their circumstances.

CHECKLIST

- Members must demonstrate their commitment to EA. A special environment committee acts as an expression of priority and a mechanism to ensure progress. The audit should be a corporate initiative.
- Establish a permanent EA steering group. This should provide the focus for management. All functional areas of the authority, and other groups if appropriate, should be represented.
- If EA is to succeed it must be managed as an internal process of learning and change. To maintain this in the long term, the structures, experience and sense of 'ownership' must be built up.
- An environmental coordinator is an essential element in the management system. The post should be located so as to maximize interdepartmental cooperation.
- Consultants should be used in a supportive role, offering advice, training and objectivity.
- Generating a feeling of ownership throughout the authority is an essential task of EA. The management system should embrace all levels and all sectors, emphasizing the role and responsibility of each individual, through a programme of training and awareness raising.
- The public should not just be consulted on the audit but should be facilitated to participate as active members. They can provide valuable advice and information and become partners in implementing change.
- Resource efficiency should be a key goal of EA. Monitor the costs of maintaining the process and of implementing the output.

REFERENCES

Grayson, L (1992) Environmental Auditing - A Guide to Best Practice in the UK and Europe Technical Communications and British Library Science and Information Service, Letchworth.

CHAPTER 15

FUTURE PROSPECT OF ENVIRONMENT AUDITING

INTRODUCTION

This chapter acts as a conclusion of the last two chapters, reiterating some of the main points of EA processes. Two themes in particular are discussed, implementation, mechanisms and the role of individuals, because they are illustrative of a number of fundamental issues.

Also, this chapter rounds off the discussion on EA in general and the role that the process can play in the search for sustainable development. It examines the prospects for EA within an evolving system of local environmental management and then explores the role of the EA process in the pursuit of sustainable development goals.

IMPLEMENTATION MECHANISMS

Implementation mechanisms are by right an element of all other management systems. However, they are of such importance that they deserve special mention. Examining these mechanisms also emphasises the imperative of producing an implementable set of actions. This should be used as a principle to define the purpose and direction of auditing activities.

Some of the mechanisms were discussed previously in relation to form and format for action recommendations. Overriding these details, however, is the requirement for a high level commitment to action and the management mechanisms to deliver it.

On a very pragmatic level, the first real test for members is when verbal support is to be turned into financial backing. Many a far-reaching goal in EA has been circumscribed when it cam to deciding between allocations for the audit and for another interest, like education or housing (through ideally many interests should coincide). Where real choices are concerned, it will be the personal and professional commitment on the part of members that decides whether EA is favoured.

In a different sense, the same is true for officers, other staff and the wider public. An involvement throughout the process will be crucial in generating ownership, which will in turn dictate the degree of progress possible. In most cases the changes called for in the audit will impact on either the general local authority staff or the wider public. It is also here that the responsibility for implementing most of the actions will lie, rather than with high ranking officers or members. It is therefore important that potential resistance to change is avoided through involvement and awareness raising. The public's role in implementation should also be explicitly addressed. Often a dual set of mechanisms will have to be set up; internally a regime will be established in order to make the action happen externally. So within the authority, service and personnel development plans might need amending: new budget allocations made; a reporting structure put in place, etc... These mechanisms are then used to facilitate a better relationship with the public, through for example cooperation, education and grant provision.

There are two final points on implementation mechanisms. Firstly they must be dispersed throughout the authority in order to encompass even those not directly involved in the earlier EA stages. Secondly, they have an external component, broadly defined as community involvement, which must be managed alongside the internal regime.

INDIVIDUALS IN ENVIRONMENTAL AUDITING

The discussion so far has been presented in a reasonably objective and rational way. The points have been discussed earlier. In actual fact the EA process is neither totally objective nor totally rational. The reason for this is that individuals are involved, and this in turn means that values, perceptions, beliefs and feelings, together with the whole gambit of other human factors, influence the process. Despite many people's efforts these will never be comprehensively understood, either in relation to the environment or any other matter.

While human factors can not be entirely planned for, a realisation of their significance is extremely important. This requires some angle from which they can be looked at; that is a factor, which reveals the underlying beliefs but is more open to examination. Commitment - the willingness to provide high level support on an on-going basis - is such a factor. The importance of commitment in all stages of the process lies in the fact that EA is not yet institutionalised within local government. While the legislative mandate, policy context, management structures and procedural methods are still developing; the institutional gap (as such) is being filled by the commitment of key councillors and officers. Practically every authority encountered, certainly those who have made significant progress, the influence of these key individuals has been profound. Box 16.1 presents two cameos to give an indication of the difference that personal commitment can make to the process.

The goal of EA should be to develop and spread this sense of commitment so that as the institutions supporting the process grow in strength so too does the level and depth of personal concern. This will simultaneously create the opportunity and the willingness to act.

PERSONAL COMMITMENT TO EA

- Should be personal and political
- Should be well aware on the issues of environment
- Must play a key influensive role in establishing/ setting up environmental strategy
- Set driver to process forward
- Lobbying other politicians/councillors support and encourage-ment

PROSPECTS FOR THE ENVIRONMENTAL AUDITING PROCESS

It has been emphasised throughout this book that EA is not a limited, one-off product but rather a strategic and evolutionary process. The elements in and management of the strategic process have been discussed at length. This section explores the prospects fcr the future development of EA. It should be read in conjunction with Appendix, which highlights some of the main contextual factors likely to influence local authority action.

The role of coordination

This book has attempted to define EA very broadly because it is apparent that the process has now become not just one, buy a whole set of overlapping and interlinked activities. These were described initially in beginning chapter as the tasks of MA, RIP, PIA and SoE. However, whole sets of related activities have also been touched on. These have included environmental management systems, strategic environmental assessment. These emerging activities are in one sense independent processes designed to fulfil their own sets of goals. From another perspective, however, they are all intimately connected within an extended system of environmental management. Ultimately they all set out to review environmental performance and to recommend ways whereby human impact can be reduced. This is exactly the rationale for E.A. So, in theory there is much overlap between these activities. Exactly what is happening in the real local government setting is more difficult to say. Experience to date would appear to confirm a high degree of interrelatedness between the activities and perhaps more importantly between the management of the activities. A couple of examples of this are the expanding role of environmental coordinators, originally employed for EA but now also developing Local Agenda 21, and the use of PIA methodologies in the environmental appraisal of development plans.

Figure (the environmental auditing process) shows the overlapping system of activities in this expanded view. This figure also demonstrates the pivotal role that EA has within this system. It forms the hub around which much of the activity circulates. Clearly, the problem that could easily arise from this conception is that so much is sucked into the centre of the system that it becomes impossible to manage and coordinate all the separate activities. While this is potentially a grave problem, the solution actually lies within EA. As originally conceived, an audit was not only designed to review environmental performance but also to coordinate the response to the problems that were found. The importance of this coordinating role has not exactly been lost along the way but it has equally not been highlighted in the majority of audits. Faced with the realities of traditional bureaucratic systems, auditors have often had no choice but to assess issues separately and to pursue sectoral responses. The danger with this type of approach has always been that problems are shifted rather than solved and that the essential areas of overlap are forgotten. In the long run progress will no doubt be made but it will involve more time and effort than if a coordinated approach was followed.

The introduction of environmental management systems

A revival of the coordinating role of EA is no doubt set to come about with the introduction of environmental management systems. EMAS in particular lays great emphasis on this with not only an overall management system for each of the registered operational units but also a 'Corporate Overview and Coordination System' to ensure that the audit is the responsibility of the entire authority.

With a management system as a core element in an expanded view of the EA process, each of the activities would be tied together and brought forward as a unified whole. Each would also be working to achieve the principles and goals set out in an environmental statement. If this statement were based on a vision of sustainable development then the separate activities would be seeking to integrate different aspects of the vision. What this involves is a view of sustainable development that is both far-reaching and practically useful.

MANAGING THE CHANGE TO SUSTAINABLE DEVELOPMENT

Some of the political realities of managing the change to sustainable development are necessary to discuss. It is argued that environmental auditing provides a process whereby a local authority, with the support of its community, can effectively implement the desired changes. In this respect it has been demonstrated that EA is sufficiently synoptic to encompass the relevant issues, is realistic enough to be accommodated by local authority organisations and is capable of directing strategic action within a flexible, honest and transparent system.

While the intrinsic value of EA is hopefully no longer a question, there remains the outstanding issue of the value that society places on such tools of environmental management and it relates to the relative importance that governments and the international community place on environmental factors in the conduct of their affairs. It can be argued that while environmental issues are now receiving more attention than in the past they are still not a priority when compared to the more established concerns of economy, defence, health, education etc. Witness, for example, the almost total absence of environmental safeguards in the recent revisions to GATT or, more locally, the absence of environmental concerns in current debates about the national curriculum. Sustainable development has done much to align environmental issues alongside other interests but as yet has done little to shift the balance of interests. The intractability of established systems to change is therefore a factor, which needs to receive special attention.

In the local government context there may be two aspects of the established system that present a particular challenge: the domination of economic interests and the hierarchical mode of management. There is currently a good deal of effort being put into examining the economy -environment relationship. In particular, the overseas apex bodies have recently produced a guide to local authority initiatives in this are which sets out approaches to making economic development more environmentally sustainable. On a different level of science of environmental economics is continually evolving more useful techniques and concepts which demonstrate the interdependence of the two components. There are no easy answers when it comes to the 'greening' of economic development. It is, however essential that EA asks the right questions.

The second challenge to EA comes from the system of hierarchical control, which often dominates the local government

Table 15.1 Local authority perspectives on environmental concerns

Traditional environmentla perspective	*New/broad environmental perspective*	*Sustainable development perspective*
Health and safety	Environmental risk assessement	Environmental quality management
Waste collection and disposal	Waste management strategy	Equity in access to environmental resources
Pollution Control	- reduction and recycling	Closure of resource loops
- point source and single substance	Energy conservation	- pollution and waste avoidance
	Prevention of pollution	- more efficient use of resources
		- multiple purpose and use.
Land use control	Integration of landuse planning and other policy goals	Strategic environmental assessment
Heritage conservation		- hierarchy of policy, plan, programme and projet
Achieving balance and quality through development planning	Reduction of environmental impact	Unreachable environmental constraints
		Systems view of environment-economy-society relationship
Nature conservation	Habitat enhancement	Contribution to global biodiversity
- designated sites and endangered species	Consideration of total natural resource	Natural resource constraints
Open space provision for amenity		Intrinsic value placed on other species
Landscape protection		
Single issue monitoring	Multiple issue monitoring and review	Monitoring, review and feedback within holistic environmental system
Public participation	Education, advocacy and awareness raising	Community involvement
	Access to information	Access to agenda setting and decision making processes

sector. Chapter 14 has shown how these control structures are of fundamental importance in the conduct of EA. This book suggested maximising the effectiveness of the established systems and also expected to incorporate alternative approaches, if needed. It does not mean that we have suggested to change the existing pattern of control. Within established systems there may be a considerable scope to develop participatory mechanism. A wish to generate auditing ownership through effective involvement should be the answer of this question.

CONCLUSION

Environment auditing is therefore, to evaluate the present situation in terms of understanding the degree and direction of changes required in the greater interest of global, national and local environment protection and also the sustainable development efforts. It is not only a tool for environmental awareness or management but has a significant role as a process of sustainability and development. By this process, the process of change will have certainly initiated in the right direction.

Part-Three

Environment Audit in India

CHAPTER 16

ENVIRONMENT AUDITING IN INDIA

In India, environment audit was introduced as an exercise of self-assessment to minimise the generation of wastes and pollution control. A gazette notification was issued, in this regard, by the Ministry of Environment and Forests on March 13, 1992 and later amended vide notification GSR 386 (E) on dated April 22 1993. This notification applies to every person carrying on an industry, operation or process requiring consent to operate under Section 25 of the Water (Prevention and Control of Pollution) Act, 1974 (6 of 1974) or under Section 21 of the Air (Prevention and Control of Pollution) Act, 1981 (14 of 1981), or both, or authorisation under the Hazardous Waste (Management and Handling) Rules, 1989, issued under the Environment Protection Act, 1986 (29 of 1986). The notification requires that an Environment Statement for the financial year ending the 31st March be submitted to the concerned State Pollution Control Board, on or before the 30th September of the same year.

BACKGROUND

Environment Auditing began in the United States and United Kingdom. This early environment audits originated from commercial response to national requirements. Later, it resulted as legislation, which made companies responsible for environmental damages, caused by them. In the seventies of the last century, the US adopted in principle "The Polluter Pays". And, the companies started their Performance Review and Compliance Audit in order to avoid any liability with regard to the legislation. During 70s and 80s of the 20th century, the number of anti-pollution laws and regulations framed, in which Resources Conservation and Recovery Act (RCRA), Comprehensive Environmental Response, Compensation and Liability Act (CERCLA) and the Clean Air Act (CAA) were most significant.

The concept of commercial environmental auditing has widened and now it has become a major tool for promoting effective environmental management. The typical commercial audits involve inputs and outputs analysis of the factory and tells about the environmental impacts of the raw materials and products, the impacts of the products and wastes that emerge from the factory as a result of production and administrative processes.

The environment audit to the local authority sector began in the US soon after publication of the Friends of the Earth " Environmental Charter for Local Government" in 1989. And, it has become an essential within local authorities in terms of application of the local agenda 21 in European countries, followed by other countries.

In India, the environment auditing within local areas-other than the industries and manufacturing units has been started partly and voluntarily by some enterprises to show their concern with the environment. This is however, a small beginning but these practices certainly will take a greater shape in near future.

Environment Auditing in Local Areas

In India, a policy statement for abatement of pollution, 1992 announced by the Union government seeks integration of environment considerations in to decision making at all levels. Environment Audit has been recognised as one of the instrument for achieving this objective.

According to the policy statement " industrial concerns and local bodies should feel that they have a responsibility for abatement of pollution. The procedure of an environment statement will be introduced in local bodies, statutory authorities, and public limited companies to evaluate the effect of their policies, operations, and activities on the environment.

In the changing environmental scenario globally, a pressure for change is coming from bottom - the people as voters or consumers/end users, and also from the global community. Government, Corporate and Local Authorities of the global colony are in the dock who have to made answers of the environmental fitness. In the aftermath of the Earth Summit, Rio in the year of 1992, the Governments are getting in the Rio spirit and have produced a national strategy for sustainable development, which includes policy programmes aimed at safeguarding and promoting geo-biodiversity and stabilising green house emissions.

At the local level - in the overseas countries- every authority has been invited to begin work on the Local Agenda, 21 (especially in EU member countries) to fulfil the commitments, which was made at Rio Earth Summit. Some local authorities in the overseas countries- have already taken a well and advance step, in which some newly set up private-public partnership firms/ trusts may be underlined who have successfully implemented the local areas environment auditing. In India, it is yet to be exercised.

An environment audit within local areas does not have to cover every aspects of the environment or of the authorities, companies operations nor does it have to be undertaken all at once. Auditing is a process; it can be conducted on whatever scale is appropriate to the needs and resources of the authority concerned. Moreover it does not require the extensive use of consultants. Some training and assistance may be necessary, but in most cases it may be actually be more appropriate if the authorities own staff carries out auditing. Environmental Auditing is therefore, just an open to and useful for small districts as it is for country and city authorities. (One of the key conclusion of the Local Government management Board study on auditing)

In fact, the term " environment auditing" within local authority is broader than in the industrial and manufacturing sectors and it is a hard core fact that the sustainability can only be ensured by auditing of the environment within the local area in a greater way.

Whose Environment Audit ?

In India, environmental auditing is meant for those industrial houses who requires consent of the Government under the Water

(Prevention & Control of Pollution) Act, the Air (Prevention & Control of Pollution) Act and who require authorisation under Hazardous Wastes (Management & Handling) Rules. It is made mandatory to all the polluting industries listed with the regulatory authority. The polluting industries are now spell bound to submit a periodical environment audit report in order to meet regulatory requirements.

Definition

Environment auditing is a management tool comprising a systematic, well documented, periodic and objective evaluation of how well the management systems are performing to meet the regulatory requirements to conserve environment. In terms of polluting industries, the environment audit is aimed at

(i) to prevent and reduce wastes,
(ii) to assess compliance with regulatory requirements,
(iii) to facilitate control over the environmental practices by a company's/authority management,
(iv) and, to disseminate environmental information to the public or the users.

NEED OF ENVIRONMENTAL AUDITING IN INDIA

Pollution from our Industries

India is on the developing stage. We require developed science and technology to develop ourselves. With the growth of Science and Technology, various types of Industries come with new technology and materials aimed at to provide comfortable life but at the same time it gives more penetrating / biting pollutants to us. And, we are creating our comforts at the risk of our lives and the future generation as well.

With the enthusiastic hope of green revolution in India, we are proceeding ahead with producing more chemical (fertilisers) factories. The effect of chemicals on the soil, on our ecology and on our environment is not only adverse, it is alarming too. Chemical plants generate large quantity of air pollutants, the main effluents are fluorine gas, particulate, sulfur dioxide and trioxide from sulfuric acid or phosphoric acid units and nitrogen oxides, and ammonia, hydrocarbons and particulate from nitrogen based plants. It is unfortunate to note that although compared to the cost of plant itself, pollution control equipment's costs very small. But most of our units-whether it is of public sector or of private-has not shown their self-spirited desire to install the required equipment's.

Similarly the textile mills are producing cloth for us. Today we cover our body even our bed with fabrics produced by the textile mills. Textile mills operate in humid atmosphere and handle a variety of organic chemicals in a sizable quantity. Their major effluents are cotton dust, smoke and combustion wastes, kerosene or naptha vapours, sulphuric acid, nitrogen oxides, and chlorine formaldehyde and chlorine dioxides. The ill effects of all such pollutants have not been fully realised in our country. Areas around textile mills are usually covered with a deposit of cotton dust and fabric particles. Workers in the Textile Company, inside the mill and even residents outside have no way to be protected but to inhale it. Continuous inhalation causes by sinosis, weakens respiratory functions and resistance to lungs diseases, mainly TB and Chronic bronchitis.

The manufacturing process of basic organic chemicals generates various hazardous semi-solid residues. These residues or backends from distillation columns generated by a variety of nitro and/or chloro nitro aromatics, aniline, metaminophenol.

Aniline - colourless oily liquid used in making dye-manufacture from nitro benzene produced about 5 kg of backend per tonne of product containing 50% phenyl cyclohexy lamine, 30% orthoaminophenol and 20% terry matters besides traces of nitro benzene in aniline. Manufacture of Nitritoluene generates about 24 kg of semi-solid residue per tonne of mixed Nitrotoluene. The residue contains 40% di-nitrotoluene, 10% P- and m-Nitrotoluenes and 5% O-Nitrotoluene. The rest is made up of water smeared of black tar. Dinitrotoluene is explosive and hazardous.

We produce variety of insecticides, fumigants, herbicides, fungicides, weedicides, rodenticides, pesticides and antibiotics for

PRESENT STATUS

Statewise (as on June 30, 2000)

S. No.	Name of the State/UT	No. of defaulters	No. of Industries Closed	No. of Industries which have provided requisite treatment/ disposal facilities after issuance of directions	No. of defaulters
1.	Andhra Pradesh	60	17	36	07
2.	Arunanchal Pradesh	00	00	00	00
3.	Assam	07	04	00	03
4.	Bihar	14	04	10	00
5.	Goa	00	00	00	00
6.	Gujarat	17	03	14	00
7.	Haryana	21	05	12	04
8.	Himachal Pradesh	00	00	00	00
9.	Jammu & Kashmir	00	00	00	00
10.	Karnataka	20	02	16	02
11.	Kerala	36	04	32	04
12.	Madhya Pradesh	02	00	00	02
13.	Maharashtra	06	03	01	02
14.	Manipur	00	00	00	00
15.	Meghalaya	00	00	00	00
16.	Mizoram	0	00	00	00
17.	Nagaland	00	00	00	00
18.	Orissa	09	01	03.	05
19.	Pondicherry	04	00	04	00
20.	Punjab	18	01	16	01
21.	Rajasthan	00	00	00	00
22.	Sikkim	00	00	00	00
23.	Tamil Nadu	366	118	248	00
24.	Tripura	00	00	00	00
25.	UT-Andman & Nicobar	00	00	00	00
26.	UT-Chandigarh	00	00	00	00
27.	UT-Daman &Diu. Dadra &Nagar Haveli	00	00	00	00
28.	Delhi	CSP	-	-	-
29.	UT-Lakshadeep	00	00	00	00
30.	Uttar Pradesh	241	59	175	07
31.	West Bengal	30	07	22	01
	Total	851	228	589	34

agriculture and domestic uses. Apart from polluting our environment when we frequently use the insects, pests, etc. become immune to these chemicals and it can not be controlled in near future if we do not take appropriate measures to prevent. Besides the impact of manufacturing process of these chemicals on environment is also terribly damaging. When we manufacture various type of benzene hexa chloride including gama-benzene hexa chloride, near about 30 cubic meter of waste water is released per tonne of BHC. The characteristics of the waste water indicate that it is highly corrosive (3.8 kg H_2SO_4/tonne) and toxic (1.4 kg BHC/tonne). Other insecticide, dichlorodiphenyl trichlorocthane (DDT) generated 50 to 60 cubic metre of waste water per tonne of product, which is corrosive. These are only a few examples, out of various such chemicals we produce such as malathion, parathion, hexa chloro cyclo hexane, carbamates etc., all of them gives us sufficient toxic effluents during manufacturing and causes damage to our environment and habitat.

More than hundred organic chemicals, heavy metals like Zn, Pb, Cr, Cu, Hg, He, Ba and their salts are used in manufacturing of dyes and pigments. Most of these are discharged into waste water besides the small quantities of intermediate compounds along with final products. Some of them are highly toxic. These refuses are thrown on fallow land within the vicinity of manufacturing units and accumulated this with rains. This may create toxic substances to sub-soil water.

Caustic chlorine industry is another manufacturing area which gives us enough pollutants to poison our life and life sources. It is estimated that the rate of mercury consumption is 394 GM per tone of caustic against the targeted value of 350 GM of mercury per tonne while it is 90 gm of mercury per tone of caustic in industrially developed countries. Mercury is lost in the different routes such as solid wastes and sludge's, water, product hydrogen and caustic, handling loss and unknown loss, which includes loss of mercury as vapour. This loss of mercury pollutes our environment and water with the expectation of creating poison to us through fishes.

The tannery effluent is playing havoc with our environment and if it allowed to function as it is today, we shall have to live with polluted sources. The paper mills all over the country are polluting through its liquid disposals and gas-solid as well.

In fact, every industrial unit in our country gives us more pollution than the products by way of excess usage of the raw materials. These excess usage of raw materials cause wastes and thus, pollution on environment. Due to complexed problems related to waste generation, its quantity and characteristics-it becomes an ineffective and uneconomical. The waste generation varies time to time or seasonally, even it may vary hourly or daily depends upon the multiplicity of manufacturing product at the same place. The wastewater characteristics also largely vary from stream to stream discharged from industrial unit operations of a particular product. In this uncertain and growing complexity problems, the possible way of waste prevention and reduction workout to be more effective.

It is essential to check whether an industry or a company is complying with the environmental norms / standards alongwith other regulatory requirements. It is also important to monitor these aspects periodically, which include gaps determination and workout of action plans for implementing within retention time. In case the gaps an identified for compliance in the light of the regulatory requirements, the regulatory body could be apprised of these action plans and retention time. Finally, the regulatory risk could be overcome and corrective and effective steps will be taken for pollution control.

Advantages of the Environment Audit in Industries

Environment auditing has far reaching benefits to the industry, to the society and the nation at large. Thus, the benefits of the Environmental Auditing in terms of Industry or company, can be underlined as follows:

(i) Determines how well the process systems and pollution control systems are performing,
(ii) Identifies the poor performances of the systems/operations,
(iii) Identifies potential cost savings by way of waste minimisation, adoption of waste recycle / waste reduction / waste recovery methods,

(iv) Provides up to date data base on environment for future use in plant modification or emergencies,

(v) Helps in understanding the technical capabilities of man and machine and attitude of the environment people in a company or industry,

(vi) Probe and solve surprises and hidden liabilities, due to which regulatory risk and exposure to litigation can be minimised / reduced,

(vii) Identifies environmental issues that requires attention of the company/industry, warns timely to the management on future environmental problems, encourage and ensure independent verification,

(viii) Assists in complying with national, regional, and local laws and regulations on environment with the company's policy and with the environmental standards,

(ix) Increases awareness on environmental requirements, policies and responsibilities

(x) Helps in safeguarding the environment.

Objective view on the Environment Audit

The environment audit helps in controlling pollution, improving production, measuring safety and health status, conserving environment and managing natural resources and overall in achieving sustainable development. In an industry, the objectives of environment audit can be stated as follows:

(a) To determine the various materials used during manufacturing process and the performance of various process equipments,

(b) To identify the usage of raw materials in excess than required,

(c) To review the conversion efficiencies of process equipment and accordingly fix up norms for equipment/operational performance and minimisation of wastes.

(d) To identify the areas of water usage and waste water generation,

(e) To determine the characteristics of waste water,

(f) To determine the effluents, emissions - its sources, quantities and characteristics,

(g) To determine the solid wastes and hazardous wastes - its sources, quantities and characteristics,

(h) To identify the possibilities of waste minimization, recycling of wastes or recovery of wastes,

(i) To determine the impact on the surrounding environment (groundwater, stream, residential area, agricultural area, sensitive zone, etc.) due to the disposal of wastewater, emissions and solid wastes from the industry and accordingly identify suitable preventive measures, if necessary;

(j) To verify compliance with the standards and conditions prescribed by the regulatory bodies under the Water Act, the Air Act and the Environmental (Protection) Act; and

(k) To check the effectiveness of (a) organisational set-up of the industry for decision-making and environmental management with special reference to their 'technical' view point, 'attitudinal' viewpoint and training, and (b) environmental policy of the Company.

Introduction

The Government of India has been increasingly concerned about the control of environmental pollution specially due to industrial activities. This is evident from the pollution control legislation enacted by the Parliament and follow-up programmes for their imple- mentation. These programmes involve three different approaches, namely,

(i) tackling of the pollutants;

(ii) tackling of the polluted areas;

(iii) tackling of the polluting sources. Direct control of the pollutants includes the reduction of lead content in motor spirit, controlling mercury pollution from caustic soda industries, improved house-keeping for controlling discharge of heavy metals, like chromium and nickel, in electroplating industries etc.

Controlling polluted areas necessitates an integrated approach towards environmental management through control at source, which in turn involves concerted efforts in evolving time-targeted action plans, and their implementation through various agencies concerned. The third approach involves securing compliance with the effluent/emission standards prescribed in respect of the polluting industries.

The Central Board has been actively involved in developing the sectorwise standards at national level, for effluents and emissions from different polluting industrial sectors, and formulating nation-wide programmes for their effective implementation. The State Pollution Control Boards (SPCBs) have been persuading the industries since the enactment of the Water & Air Acts and rules thereof to make them comply with the standards. In addition to this, National level programmes for control of discharges/emissions from polluting industries have also been taken up. The details of these programmes and the status of implementation of these programmes are presented in the following sections.

Polluting Industries

There are 64 types (listed below) of polluting industries/ industrial activities, which are classified as "Red Category" industries on the basis of their emissions/discharges of high/significant polluting potential or generating hazardous wastes. These include large, medium as well as small scale industries.

Industries identified by Ministry of Environment & Forests, Govt. of India as heavily Polluting and covered under Central Action Plan, viz.;

1. Distillery including Fermenta tion Industry
2. Sugar (excluding Khandsari)
3. Fertiliser
4. Pulp & Paper (Paper manufacturing with or without pulping)
5. Chlor Alkali
6. Pharmaceuticals (Basic) (excluding Formulation)
7. Dyes and Dye Intermediates
8. Pesticides (Technical) (excluding Formulation)
9. Oil Refinery (Mineral oil or Petro refineries)
10. Tanneries
11. Petrochemicals (manufacture of and not merely use of raw material)
12. Cement
13. Thermal Power Plants
14. Iron & Steel (Involving processes from ore/scrap, and Integrated Steel Plants)
15. Zinc Smelter

16. Copper Smelter
17. Aluminum Smelter

Industries manufacturing following products or carrying out following activities:

18. Tyres and Tubes Vulcanisation/ Retreading/molding
19. Synthetic rubber
20. Glass and fibre glass production and processing
21. Industrial carbon including electrodes and graphite blocks, activated carbon, carbon black etc.
22. Paints and Varnishes (excluding blending/mixing)
23. Pigments and intermediates
24. Synthetic resins
25. Petroleum products involving storage, transfer or processing
26. Lubricating oils, greases or petroleum-based products
27. Synthetic fibre including rayon, tyre cord, polyester filament yarn
28. Surgical and medical products involving prophylactics and latex.
29. Synthetic detergent and soap
30. Photographic films and chemicals
31. Chemical, petrochemical and electrochemicals including manufacture of acids such as Sulphuric Acid, Nitric Acid, Phosphoric Acid etc.
32. Industrial or inorganic gases
33. Chlorates, perchlorates and peroxides
34. Glue and gelatine
35. Yarn and Textile processing involving scouring, bleaching, dyeing, printing or any effluent/emission generating process
36. Vegetable oils including solvent extracted oils, Hydrogenated oils.
37. Industry or process involving metal treatment or process such as pickling, surface coating, paint baking, paint stripping, heat treatment, phosphating or finishing etc.
38. Industry or process involving electroplating operations
39. Asbestos and asbestos based industries
40. Slaughter houses and meat processing industries
41. Fermentation industry including manufacture of yeast, beer etc.

42. Steel and steel products including coke plants involving use of any of the equipment's such as blast furnaces, open hearth furnace, induction furnace or an arcfurnace etc. or any of the operations or processes such as heat treatment, acid pickling, rolling or galvanising etc.
43. Incineration plants
44. Power generating plants (excluding D.G. Sets)
45. Lime manufacturing
46. Tobacco products including cigarettes and tobacco processing
47. Dry coal processing/Mineral processing industries like ore sintering, palletisation etc.
48. Phosphate rock processing plants
49. Coke making, coal liquefaction, coaltar distillation or fuel gas making
50. Phosphorous and its compounds
51. Explosives including detonators, fuses etc.
52. Fire crackers
53. Processes involving chlorinated hydrocarbons
54. Chlorine, Fluorine, bromine, iodine and their compounds
55. Hydrocyanic acid and its derivatives
56. Milk processing and dairy products (Integrated Project)
57. Industry or process involving foundry operations
58. Potable alcohol (IMFL) by blending or distillation of alcohol
59. Anodizing
60. Ceramic/refractories
61. Lead processing and battery reconditioning & manufacturing including lead smelting
62. Hot mix plants
63. Hospitals
64. Mining and ore-benificiation

CHAPTER 17

ENVIRONMENT AUDIT PROCEDURE

The audit procedure includes:

(1) Pre-audit activities.
(2) Activities at the site; and
(3) Post-audit activities.

The details of these activities and the entire environmental audit procedure are shown in Fig. 1 under caption " environment audit procedure".

PRE-AUDIT ACTIVITIES

Preliminary information

Pre-audit activities include various preparatory works. Having chosen the industry to be audited, preliminary information on the industry is to be obtained through a questionnaire. The information include location of the industry with surrounding land uses, gaseous emissions, solid waste/hazardous waste, and organisational set-up and policies of the Company for environmental management. A typical questionnaire is given in annexure II.

The preliminary information received on the industry should be reviewed to identify main areas of concern. Thereafter it is required to prepare and organise audit team and resources, and allocate specific tasks to team members. Resources such as the sampling and monitoring equipment and laboratory facilities for analysis should be checked if available at site or else arrangements should be made for their availability through external sources such as private/government laboratories, loan from other industries etc. The visit programme should then be intimated to the industry mentioning that the environmental audit should not be considered as a raid. The prior intimation to the industry helps them convince the senior management and staff at various level of the purpose of audit and the cooperation they have to extend to the audit team. The staff should not feel that the audit would lead to surfacing problems and hence they would be subject to criticism by the management. They should be made clear about the purpose and objectives of the audit and how beneficial it would be for the industry. This would also increase employees' awareness towards waste reduction and promote input and support for the audit.

Audit team

Audit team should be carefully selected to cover various aspects of the audit. The team should be employees from production, quality control laboratory, R&D, pollution control operations, technical staff for monitoring and analysis of waste samples and environment and an environment specialist. The number of people may vary from 4 to 8 depending on the size and complexity of the facility being audited.

The team should be sufficiently detached to provide an independent view. The members should be so chosen that they would not hesitate bringing out even criticism, owing to obligations with supervisor. Sometimes it is advantageous to include members from the headquarters of the industry.

Effectiveness of audit is a direct result of the qualification, confidence, training and proficiency of the personnel who conduct audits. The team should understand regulatory requirements, relevant waste control technologies and their operations and process. They should have capability to examine, question, sample and analyse waste and interpret data. The management should be provided with a realistic assessment of environmental performance.

It is important to have well-defined and systematic procedures which are known and understood by all concerned. The audit may take 3-10 days depending on the industry's size and products.

Effectiveness of audit is a direct result of the qualification, confidence, training and proficiency of the personnel who conduct audits. The team should understand regulatory requirements, relevant waste control technologies and three operations and process. They should have capability to examine, question, sample and analyze waste and interpret data. The management should be provided with a realistic assessment of environmental performance.

Activities

The activities at site include deriving material balance, identifying waste flow lines, monitoring of characteristics, evaluating performance of pollution control equipment/systems, assessing environmental qualities, holding discussions with various cross-sections of the staff engaged in production, laboratory/quality control, R&D, environment management, etc., so as to understand different operational mechanisms. Having a fair idea on the manufacturing process, reconnaissance surveys should be made to be familiar with layout of the plant and process operations, and to understand possible impact on the surrounding environment. Various activities to be carried out at site are discussed in details in the following paragraphs.

Material balance

The entire manufacturing process of each product should be drawn into a process flow sheet representing various unit operations as block. A unit operation is a process where materials are input, a functior occurs and material are output mostly in a different form, state or composition. A typical process flow diagram of monocrotophos, an organophosphorus pesticide is given in Fig. 3.2 under the caption " Process flow sheet of monocrotophos". This process includes the unit operations of dehydration, adduct formation, chlorination of adducts dissociation, concentration, toxification, pre-concentration and purification. A typical unit operation with inputs of raw materials, catalyst, water/air, power and recycled material and outputs of products and by-products, wastewater, emissions, solid waste and reusable waste in another operation is schematically shown in Fig. 3 under caption " a typical unit operation".

The quantities of inputs and outputs at each unit operation should be worked out for the entire process and data incorporated in the process flow sheet. Discussions with the staff, perusal of the records of the Company and the reconnaissance survey will help in arriving at these flow sheets. From these flow sheets, data sheets incorporating the raw material requirement, water consumption, wastewater and solid waste generation, and gaseous emissions should be worked out for each product manufactured. A typical data sheet is shown in Table.1 under caption "datasheet" on inputs and outputs for monocrotophos manufacture".

The water balance sheet, which shows areas of water usage and wastewater generation and their quantities, is depicted in Figure 5 under caption "mass balance of water consumption and effluent generation".

Waste flow

From the material balance, the sources and quantities of generation of wastewater, gaseous emissions and solid waste should be identified. The waste pretreatment, final treatment and disposal path should be identified. The production staff should be consulted as these people are likely to know about waste discharge points and about unplanned waste generation such as spills, and other material handling practices that are not accounted and recorded. The Quantities and sources should be accordingly finalised and a waste flow sheet prepared. A schematic waste water flow sheet is shown in Fig. 5 under caption " waste water flow lines".

Monitoring

The characteristics of the wastes as generated from the sources are important to understand its use for recycle, recovery or treatment. Also the performance of the treatment facilities are to be monitored so as to check their efficiencies and to modify or install additional equipment/facility, if necessary. The surrounding environment-groundwater, stream, soil, surrounding land uses-residential, agricultural etc., and ambient air quality should be monitored to determine the impact due to the industry. With the above objectives, sampling points should be identified and monitoring network established. Parameters to be analysed should be determined from the material balances of the wastes generated.

The frequency of sampling should be fixed so as to cover at least one full cycle of operations. More than one such set of data can result in more realistic results. Samples collected should be of 'grab' type where characteristics do not vary significantly and of composite type where characteristics fluctuate. Grab sampling means collection of sample in one pick while composite sampling requires collection of sample continuously or at predetermined frequency (such as 1 hour, 2 hour etc.) and compositioning it in proportion to the flow rate observed at each sampling time. The method of analysis of samples should be done as per standard procedure and by trained analysts.

Field Observations

The entire plant should be inspected thoroughly. The aspects of site layout, material handling and storage, drainage system, safety aspects, lapses/negligence in operations and attitude of operators in process and waste treatment facilities, handling of scrap and wastes, usage of sign boards, instructions, colour codes etc. should also be observed.

The attitude and technical capability of various staff including senior management should be observed as is very critical in achieving the goal of safe environment. The training requirements can be assessed on the basis of these observations.

Draft Report

On completing the aforementioned activities which include material balance, waste flow identification, monitoring and analysis of various samples and field observations, a draft report should be prepared with findings and possible recommendations.

The draft report should be presented before the senior management and various points should be thoroughly discussed. The management should put forward their views. The participation of the management and their acceptance on various observations and recommendations makes the tasks meaningful.

POST AUDIT ACTIVITIES

Synthesis of data

The requirements of different raw material according to the mass balance of chemical equation involved in the manufacture of a product are called stoichiometric requirement. A comparison of these requirements with the actually used in the industry gives an indication of excess usage of various raw materials. These excesses may be presumed to be finding their way to air, water and soil thus causing pollution. Hence, it is important to reduce these excesses. The unit operation should be checked up to find out the cause of excess usage of the materials and accordingly modifications made. Norms should then be fixed for performance of each of the unit operations, for wastes generated from each of these unit operations. The production and environment staff is simply to adhere to the norms. The Environment Manager thus can have a control over production as well as wastes generation too.

Evaluation of waste treatment facilities

Performance of various pretreatment and final treatment facilities should be evaluated based on the analysis report. If the treated wastewater, gaseous emissions and solid waste do not conform to the standards prescribed by the Pollution Control Board, reasons for the same should be diagnosed.

From the individual streams of wastewater, recyclable and recoverable materials should be identified and provisions made for the same. All the 'avoidable' allowed for discharge. The wastewater should be segregated based on the characteristics, such as inorganic, organic, acidic, alkaline, easily biodegradable, not easily biodegradable and toxic streams; and pretreatment units viz. oil separator, neutralisation. detoxification etc. should be provided, wherever required, at the source so as to minimise cost of final treatment.

The wastewater of similar nature is combined and common treatment facilities provided. This would be efficient and economical. Many a times. It is observed that inorganic wastes and non-biodegradable wastes are treated in biological treatment plants, which on the contrary render biological treatment ineffective. Toxic wastes should be detoxified before treating in biological treatment plant. Highly toxic wastes may be isolated and incinerated. The rate of wastewater flow and polluted loads to the effluent treatment plant (ETP) should be properly regulated to keep off shock loads to microorganisms. The designed criteria and the actual operating conditions of various treatment units should be compared and norms fixed for the operation of these units.

Similarly, the problems related to gaseous emission and solid waste generation may be identified. Recommendations for the best practicable waste management systems should be formulated. The guidelines for environmentally safe layout are given in Annexure III and guidelines for reduction of raw materials losses, and wastewater and gaseous emissions are given in Annexure IV.

The Environment Division of the industry should have an environment specialist out look into matters related to pollution control and evolve norms for resource conservation/waste minimisation vis-a-vis process control. Besides, he should also evolve norms for optimal utilisation of resources and performance of various and pollution control systems. The members of this division and the operators of the treatment facilities should be well trained.

To oversee the implementation of the measures for pollution control and the overall management of environment, there should be a Peer Group comprising members from production, quality control/laboratory, R&D and waste treatment divisions, the top management, and an environment specialist.

Final report

Various aspects discussed above should be compiled and a final report prepared alongwith recommendations. The final report may, if necessary, be sent to the top management for comments so as to make further modifications.

Action plans

The recommendations include measures for best practicable environmental management. If the annual burden. i.e. the annualised capital cost of the pollution control measures and their operating cost, for the implementation of all the recommendations, is high and the investment not feasible for the industry, then these recommendations, should be implemented in phases. Priorities should be fixed and action plans with time-frame should be formulate.

Follow-up actions

Follow-up actions should be taken to check the progress of implementation of recommendations. The Environment Division of the industry should meet the other divisional heads periodically to review the progress.

CHAPTER 18

ENVIRONMENT AUDITING : WHAT, WHY AND HOW

ENVIRONMENT AUDITING WHAT? WHY? HOW?

WHAT IS ENVIRONMENTAL AUDITING ?

Environmental auditing is a management tool comprising a systematic, documented, purist, periodic and objective evaluation, who shows how well environmentally concerned organisations, management and equipment's are performing as safeguard of the environment and with the aim regulate the environment by facilitating management control over environmental practices; and by assessing compliance with company policies which would include meeting regulatory requirements.

In our country, presently the environment auditing is confined to industrial/manufacturing organisations, therefore, we may define it as a management tool for systematic and .periodic evaluation of manufacturing processes viz-a viz waste generation and their generation. In fact, it is an essential ingredient of total quality management.

Environment audit can be sub-classified in following categories;

- Waste audit
- Energy audit
- Health and safety audit

RESOURCES OF ENVIRONMENT AUDIT

- **Benefits** of environment audit are multi-dimensional. It optimises the manufacturing process and provides assistance to quality assurance from raw material to product. Besides this, it minimises hazard risk, wastes generation, and above all helps in conservation of material.
- **Water management** - water is a limiting factor for industrial activities. Its management is, therefore, pre-requisite for sustaining and optimising industrial processes. Recycling, recovery, renovation, and reuse of industrial wastewater require engineering innovations.
- **Material and Energy Balance** - In any chemical process, loss from material is inevitable. Obsolete technology often adds to material loss to such an extent that the production becomes not viable. It damages the environment also. Environment audit can help in identifying the nature and extent of losses in the form of material as well as energy.

Basic steps in environmental audit

Environmental audit involves three steps, e.g. pre-audit activities, site activities and post audit activities. In case the pre-audit activities, the major task is to organise and scheduling the audit procedures. In case the activities at site, the task is to ascertain the facts. The post-audit activity is for translating the data into meaningful information for a better decision support system.

WHY ENVIRONMENTAL AUDIT ?

- Requirements of regulatory bodies
- Increasing productivity
- Financial incentives through pollution control
- Resource conservation
- Protecting against legal liability insurance act, 1991
- Assistance needed by the management for
- Environment Management Plan
- Expansion or diversification
- Requirement for insurance

BENEFITS OF ENVIRONMENTAL AUDIT

- Reduction in waste generation
- Resource conservation and therefore reduction in input cost
- Reduction in pollution potential
- Improvement in process efficiency
- Improvement in working conditions
- Assurance to staff-management-investors
- Insures regulatory authorities and general public
- Cost saving through recycling and reuse

SIMILARITY IN ENVIRONMENT AUDIT & FINANCIAL AUDIT

- Both are an activity of verification of records
- A comparison of outputs with expectations
- An assessment of unavoidable errors and waste
- Environmental audit incorporates additionally
- An assessment of risks
- An investigation of plant performance
- An assessment of material and energy input and output including possibilities of waste minimisation

ENVIRONMENT AUDITING OBSERVES

- What the management or Authority is doing?
- Can the management or authority do it better?
- Can the management/authority do it more economically?
- What more the management / authority should do?

ENVIRONMENT PROTECTION COST

DIRECT COST	INDIRECT COST	LIABILITY COST
CAPITAL EXPENDITURE	ADMINISTRATIVE COST	PENALTIES
BUILDING UTILITY	REGULATIONS/COMPLIANCE COST	FINES
PROJECT ENGINEERING	INSURANCE	EXPOSURE & HEALTH CLAIMS
OPERATION & MAINTENANCE	WORKERS COMPENSATION	PERSONAL INJURY

HOW TO ORGANISE AN ENVIRONMENT AUDIT

- PRE-AUDIT PHASE
- NOMINATION OF THE TEAM
- SITING OUT TASKS & PRIORITY
- PREPARE BACKGROUND NOTE
- ARRANGING INTERNAL-EXTERNAL RESOURCES
- AT SITE PHASE
- INTERACTION WITH LOCAL STAFF
- FIELD INSPECTION
- SAMPLING & TESTS
- REVIEW OF RECORDS
- TENTATIVE FINDINGS
- POST AUDIT PHASE
- FINALISATION OF REPORT
- EVOLVE ACTION PLAN

IMPLEMENTING OPTIONS

- INSTALLATION-DEMONSTRATION-FOLLOW UP
- MEASURING WASTE REDUCTION
- ONGOING WASTE MANAGEMENT PROGRAMME

ENVIRONMENTAL MANAGEMENT

- PROPER RESOURCES USE AND MANAGEMENT
- AN INTERDISCIPLINARY APPROACH TO RESOURCE CONSERVATION AND RECYCLING
- ACTS AS A REGULATORY FORCE ON HUMAN DISCIPLINE IN RESOURCE EXPLOITATION AND RESOURCE WASTING

APPLICABILITY

- Those who require consent under
  Water (Prevention and control of pollution) Act, 1974
- Those who require consent under

Air (prevention and control of pollution) Act, 1981

- Those who require authorisation under
  Hazardous wastes (management and handling) Rules. 1989

CHAPTER 19

ENVIRONMENTAL AUDITING : AN OVERVIEW

INTRODUCTION

History and Definition

Environmental health and safety auditing dates back to the early 1970s when some companies developed audit programs on their own, for reviewing and evaluating environmental problems associated with their operations. This concept got projected under the guise of a number of different approaches and names such as environmental reviews, survey assessments and quality controls, environmental diagnostic studies etc., depending upon the audit programmes of the company concerned. There has been a steady growth in this discipline since then. Today the term 'Environmental Auditing (EA)'s is usually accepted and several countries have in fact formalised such programmes designed to ensure that industrial operations. Comply with the established environmental audits. However, the definition adopted by the International Chamber of Commerce (ICC) is as follows:

"A management tool comprising a systematic, documented, periodic and objective evaluation of how well environmental organisation, management systems and equipment's are performing with the aim of (i) facilitating management control of environmental practices, and (ii) compliance with company policies, including meeting regulatory requirements. It may be noted that the process of auditing as seen here is an internal process that should become a necessary and routine part of most, if not all industrial management, irrespective of the size of the company".

INTRODUCTION

Industrial pollution in our country is on increase and is creating a high-risk environment. Various legislation viz. The water (Prevention & controls of pollution) Act, 1974, the Air (Prevention & Control of Pollution) Act, 1981 and environment (Protection) act, 1986 have come into force and organisations created to combat pollution. Gone are the days when industrialisation meant profit making and environment was grossly neglected. It is being realised that industry and environment should go hand-in-hand so as to achieve sustainable development. Also over the years awareness has brought in realisation to consider environmental protection a bare necessity. Yet, the investments for such a protection are still considered a liability by many an industrialists mainly due to lack of up-to-date scientific practices of environmental management. Consideration of environmental factors at par with production helps in minimising material losses and also in reduction of liabilities in the long run.

The growing environmental pollution and the complexity of this problem with increasing risks from the regulatory controls needs and effective management tool so as to prevent pollution and to make pollution control programmes cost-effective and feasible.

'Environmental audit' is a technique being introduced for integrating the interest of the industry and the environment so that these could be mutually supportive. This technique is basically a part of industry's internal procedures in meeting their responsibilities towards better environment. Also the policy statement for abatement of pollution by the Government of India provides for submission of environmental statement by all concerned industries, which would subsequently evolve into an Rules, 1986 has been issued on April 22, 1993, requiring industries to submit an environmental statement for the financial year ending on March 31 in Form V to the concerned State Pollution Control Boards on or before September 30 every year beginning 1993 (Annexure I). The Department of Company Affairs also agreed to include this requirement as part of the Director's Annual Report. The submission of an environmental statement is applicable to the following:

(i) Those who require consent under the Water (prevention & Control of Pollution) Act, 1974.

(ii) Those who require consent under the Air (Prevention & Control of Pollution) Act, 1981: and

(iii) Those who require authorisation under Hazardous wastes (Management & Handling) Rules, 1989.

The present document outlines the guidelines for environmental audit with particular reference to pesticides industry.

PHILOSOPHY OF ENVIRONMENTAL AUDIT

Definition

Environmental auditing is a management tool comprising a systematic, documented, periodic and objective evaluation of how well the management systems are performing with the aim of :

(i) Waste prevention and reduction;

(ii) Assessing compliance with regulatory requirements:

(iii) Facilitating control of environmental practices by a Company's management: and

(iv) Placing environmental information in the public domain.

In the industry, especially the chemical industries, raw material are used in excess of the stoichiometric requirements because of the limitations on practically achievable operational efficiencies and the raw material' purity. These excess usages of raw materials, unless recovered, find their way to environment causing pollution. Wastes from an industry include non-product discharges in gaseous, liquid and solid phases. End-of-the-pipe waste treatment techniques, wherein all the wastes are carried to a common facility for treatment, is proving to be ineffective and uneconomical due to the complexity of problems associated with waste generation, their quantity and characteristics. The waste generation may vary hourly, daily and seasonally, especially in case of the multiplicity of manufacturing product in the same premises. The wastewater characteristics also widely vary from stream to stream discharged from various unit operations of a particular product. In this growing complexity of problems, the concept of waste prevention and reduction can workout to be more effective?

(i) ensures independent verification, identifies matters needing attention, and provides timely warning to management on potential future problems; and

(ii) helps to safeguard environment, and assists in complying with local, regional and national laws and regulations, with the Company's policy and with the environmental standards.

It is important to find out whether an industry is complying with environmental standards and other regulatory requirements. It is also very essential to periodically monitor this aspect, determine the gaps and workout action plans for implementation with in a reasonable time frame keeping in view the financial and other consideration of a Company. In cases of gaps for compliance with the regulatory requirements, the obtained for implementation. Thus the regulatory risk could be overcome and effective steps taken for pollution control.

Many a times, the top management of a Company or an industry may not be aware of the factual situation of their industry from environmental angle. Such unknown facts form hidden liabilities more often those not expose an industry to regulatory risks. The management should be able to periodically review the environmental practices of the Company to formulate/modify the company's environmental policy accordingly.

It is also imperative that the management of a Company should have a clear picture of 'attitudes' and 'technical capabilities' of their organisational set-up for protecting environment, pollution control status, and their bounden social obligation related to environment so as to decide on the future mode of actions. Public are to be made aware of the environmental information of the company, especially to those who are shareholders, so as to build-in among them confidence.

Environmental auditing can be viewed as a 'management tool' internally, and 'liaison' externally with the public and regulatory bodies.

Need of environmental Information

The need for environmental information could be two main types namely, (i) internal, and (ii) needs are to ensure that the industry has accurate, up-to-date information at hand on the environmental performance of its construction contractors, plants, technologies, waste contractors, products, raw material and all related activities. Such a need has in fact made environmental auditing in at least some of the developed countries, a part of the industry's overall drive for quality assurance.

The growing concern for the protection of human resources is making this area, increasingly useful and interesting even to the external agencies (i.e. other than the industry)

Among these, who are seen demanding or requesting such information are

- National and international regulatory agencies and research institutions;
- Environmental crime investigation, environmental courts etc.,
- Insures and risk manager, particularly those involved in environment impairment liability insurance field;
- Employers and trade unions;- share holders;
- Local, communities and environmental officers/planners;
- Environmentalists ; and
- Ethical investment funds wanting priority investments in companies with eco-friendly products.

Purpose and Advantages of Studies

The environmental audit studies serve at least the following three basic purpose:

(i) Compilation of the complete information on the operation of the industrial facility and its potential sources of pollution through technical inspection. This inspection, which is conducted at a facility allows the activities that are programmed and entrusted to the operational branches to be carried out in order to correct the different problems detected at their source or to foresee conservation and maintenance measures need to prevent major pollution problems.

(ii) Evaluation of the conditions surrounding the industrial facility in order to estimate possible impacts which may be caused and the suggested recovery measures for such situations.

(iii) Preparation and implementation of action plans for better control of the environment, and the environmentally related industrial activities, including further developmental activities of the areas

The primary and obvious advantage of environment auditing is to help safeguard the environment and to substantiate compliance with local, regional and national law and regulations, and with the company policy and standards. There can be several other benefits, the importance of which may vary from situation to situation. These benefits include:

- Reduced exposure to litigation and regulatory risk (e.g. prosecutions, penalties, etc.); facilitating comparison and interchange of information between operations or plants;
- Increasing employer awareness of environmental policies and responsibilities;
- Identifying potential cost-savings including those resulting from waste minimisation;
- Evaluating training programmes and providing data to assist in training personnel;
- Providing an information base for use in emergencies and evaluating the effectiveness of emergency response arrangements;
- Assuring an adequate, up-to-date environmental data base for internal management awareness and decision making in relation to plant modification, new plants etc.,
- Enabling management to give credit for good environmental performance;
- Helping to assist relation with authorities by convincing them that complete and effective audits are being undertaken;
- Facilitating the abstention of insurance coverage for environmental impairment liability.

General Approach of Environmental Auditing

The general approach followed for environment audit studies cover three main phases, namely collection information, evaluation of information collected, and formulation of conclusions, including identification of aspects needing improvement. These phase cover, pre-audit preparation, and a site visit normally involving interviews with personnel and inspection of facilities and post-visit normally involving interviews with personnel and inspection of facilities and post-visit activities. A team of experts and assisting officials, generally completes the site assignments which involve gathering all the required data/information, analysing these, drawing conclusions regarding the status of the industrial facility audited with respect to specific criteria /objective, and reporting the conclusions.

These activities are conducted within a former structure in a sequence that is repeated in each location audited to provide a level of uniformity of coverage and reliability of findings. Atypical set of step involved in environment audit is illustrated in Fig.1.1 Although, programmes may very from audit to audit, the design of each programme generally makes provision for each of the activities mentioned in the above cited illustration. These activities are briefly described in this following paragraph.

Pre-Audit activities

This covers a number of activities including, selection and review of the site and audit team, development of an audit plan which defines the technical, geographic and time scope, arranging for the expected sample collection, preservation, transport and analysis, and obtaining background information on the plant, as well as the criteria to be used for evaluating the programme. Questionnaires are preferably prepared to cover maximum data collection and wide coverage of the industrial activity as well as its surroundings and the industrial area. Depending upon the need arrangements for assistance from the local environmental authorities/ laboratories may have to be also made before the team leaves for the audit studies.

Activities at the site

The site activities include identification and assessment of the management control systems, data collection and their evaluation, and reporting audit findings. The various arrangements existing for environmental management and related aspects are identified and understood. This is followed by evaluation of the arrangements with respect to the requirements in the sense of meeting the standards regulations. The data collection include almost every aspect of the facility which may be related directly or indirectly to the emissions/discharge, as well as the details regarding the location's surroundings i.e. other industries in the vicinity, weather, soil and rain-data, ground water quality in immediate vicinity and little away. Wastewater, slugs, air quality samples, are taken depending upon the type of the industrial activity and possible impacts. The data collected are collated, compiled and evaluated, and findings are finalized on the basis of the evidence collected, compiled and evaluated, some of the findings may require transportation, for analysis of the samples collected in a laboratory, before the same are finalized.

Post-audit Activities

The activities that follows after visit to the site is completed, mainly include

(i) preparation of the final report, and
(ii) development and follow-ups for implementation of a corrective action programme.

Problems encountered during the audit

Problems, which the audit team may face, are likely to vary from facility to facility. However, what could be generally expected and needs attention, are as follows:

- The prior history of the site.
- The age of the relevant equipment.
- Lack of records related to the relevant equipment
- The attitude of the concerned personnel on site towards such audit studies.
- Problems as well as responses of the concerned management for implementation the corrective measures.

The audit Programmes in India

The concept of environmental auditing in industrial facilities in India appears to have first got into meaningful discussions in the beginning of nineties. Efforts were initiated to see the practicability of this programme before it could be made mandatory. This area gained further movement with the Govt. of India's action plan for pollution control in the critical areas including heavily polluting programme of such audit studies in 18 major polluting industries (list provided in Appendix I), in 1991-92. Meanwhile a discussion paper on "Outline of Forests (MOEF) and circulated for comment. This process finally resulted into issuing of a gazette notification on March 13, 1992 through which submission of the environmental audit reports has been made mandatory. The term 'Audit report' has also been changed State Boards on or before 30th day of September every year beginning 1993. The audit studies conducted by the Central Pollution Control Board during 1991-93 and related details are presented in this report.

CHAPTER 20

AUDITING PROGRAMME IN MAJOR POLLUTING INDUSTRIES

BACKGROUND

A nation wise comprehensive programme for carrying out environmental audits in 18 major polluting industries was prepared by the Central Pollution Control Board in the beginning of the year 1991. These audit studies were not intended for completing any regulatory requirements, but with the basic philosophy that at least those industries which need priority attention in the sense of pollution control, have a feel of what it is, as well as why and how it is to be done, before the industries actually arrange to get it done on their own as a mandatory requirement.

Objective

Although, the purpose for which such audit studies are done has been dealt in details in Chapter 1, the broad objective of conducting such studies by the Central Board involved identification of industries with good pollution control systems thus identified may be used as a demonstration for beginning effective controls in other similar industrial activities.

Selection criteria for industries

The selection of units for conduction such in-depth studies was based on the following criteria:

(i) Industries which claim to be complying with the effluents/ emissions standards and having proper environmental management.

(ii) Industries adopting clean technologies or practicing recycle/reuse to minimize waste production that can be demonstrated to other non-compliant industries.

(iii) Industries that have not yet provided pollution control systems, in order to study their problems.

Methodology

Audit team members were first of all identified in respect of each of the units studied. A detailed discussion in this connection was held among the concerned officials of the Central Board's Head Quarters as well as the various Section/Zonal offices. Audit responsibilities in respect of each of the units were assigned among the various divisions/offices of the Central Board. Arrangements were also made in such a way that the people at the head office could assist to the Zonal Office team and vice-versa in conducting audit studies depending upon the proximity of site to be audited. Laboratory and monitoring facilities wherever available with the industries were also utilised. Central Board's mobile laboratories were also utilised for remote places here stationary facilities was not available.

Each audit team submitted the report of the units audited by them for a final compilation at the Central Board's Head Quarters.

Aspects covered

Although, there were common aspects as well little variations from industry to industry or category to category, the aspects that got covered during the studies are as follows:

A. ***Status of Pollution Control***

(i) Information collection on the following:

- Products and capacity;
- Consent conditions (air & water)
- Raw material consumption;
- Water consumption (Process, cooling, boiler feed, gardening etc. separately)
- Fuel and power consumption (sulfur content of fuel);
- Chemicals (including catalysts) used in the manufacturing process/treatment systems;

- Manufacturing process(es) with flow diagram;
- Implant measures to minimise pollution including recovery & recycling;
- Lay-out plan showing collection system for effluent, storm, water, sewage, position of stacks etc.;
- Details of effluent treatment plant;
- Details on air pollution control systems;
- Solid waste generation (type, quantity & disposal);
- Hazardous waste/sludge generation, storage, treatment and disposal; and
- Receiving body of water (name, flow and quality data)

(ii) Waste stream identification and measurement of their flow and characteristics also for the combined streams going to ETP and other drains for separate disposal;

(iii) Monitoring of ground water quality near effluent storage and land disposal site;

(iv) Monitoring of receiving water quality before and after discharge of effluent;

(v) Stack emissions;

- Details of stacks with respect to height and diameter, arrangement for monitoring of emission such as port hole and platform.
- Stack emission-monitoring

(vi) Monitoring of fugitive emissions, measurement of relevant parameters for air quality at boundary limit of the industry.

(vii) Monitoring of ambient air quality with respect to concerned parameters which are relevant to the industry; and

(viii) Analysis of sludge and solid waste in case of leachable and toxic constituents.

B. *Performance Study*

(i) Performance study of the effluent treatment system provided for individual sections within the complex with respect to influent and effluent characteristics and also observation of operating parameters, which are relevant.

(ii) Performance study of combined effluent treatment system each unit processes for relevant parameters measurement of flow and collection of details on design criteria.

(iii) Performance study of air pollution control system in specific cases by conducting emission measurement before and after control device(s) for the relevant parameters.

Reports of the Environmental audits Studies

Environmental Auditing reports for the respective task teams prepared each of the units studied. The reports are quite elaborate and it is not possible to present the complete details of the each audit study in this report, for obvious reasons. A format was, therefore, developed (provided in Appendix III) to prepare a brief report of each of the units studied in a uniform manner for a systematic presentation. A total of about 120 such studies were conducted during 1991-93.

Categories of highly polluting Industrial sectors

1. Aluminum smelter
2. Basic drugs and pharmaceuticals
3. Caustic Soda
4. Cement
5. Copper Smelter
6. Distillery
7. Dyes and Dye intermediate
8. Fertiliser
9. Iron and steel
10. Leather processing industries
11. Oil refineries
12. Pesticides
13. Petrochemicals
14. Pulp and paper
15. Sugar
16. Sulfuric acid
17. Thermal power plant
18. Zinc smelter

The wastewater of similar nature should be combined and common treatment facilities provided. This would be efficient and economical. Many a times, it is observed that inorganic wastes and non-biodegradable wastes are treated in biological treatment plants, which on the contrary render biological treatment ineffective.

Toxic wastes should be detoxified before treating in biological treatment plant. Highly toxic wastes may be isolated and incinerated. The rate of wastewater flow and polluted loads to the effluent treatment plant (ETP) should be properly regulated to keep off shock loads to micro-organisms. The designed criteria and the actual operating conditions of various treatment units should be compared and norms fixed for the operation of these units.

Similarly, the problems related to gaseous emission and solid waste generation may be identified. Recommendations for the best practicable waste management systems should be formulated. The guidelines for environmentally safe layout are given in Annexure III and guidelines for reduction of raw materials losses, and wastewater and gaseous emissions are given in Annexure IV.

The Environment Division of the industry should have an environment specialist to look into matters related to pollution control and evolve norms for resource conservation/waste minimisation vis-a-vis process control. Besides, he should also evolve norms for optimal utilisation of resources and performance of various pollution control systems. The members of this division and the operators of the treatment facilities should be well trained.

To oversee the implementation of the measures for pollution control and the overall management of environment, there should be a Peer Group comprising members from production, quality control/laboratory, R&D and waste treatment divisions, the top management and an environment specialist.

Final Report

Various aspects discussed above should be compiled and a final report prepared alongwith recommendations. The final report may, if necessary, be sent to the top management for comments so as to make further modifications.

Action Plans

The recommendations include measures for best practicable environmental management. If the annual burden, i.e. the annualised capital cost of the pollution control measures and their

Categorywise Distribution of Identified 1551 (large and medium) Industries

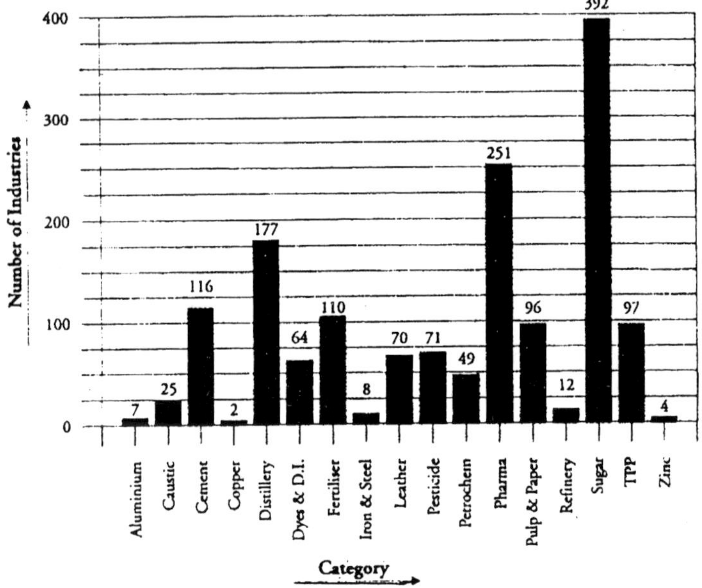

Yearwise Progress in Reduction of Number of Defaulting Industries along the Rivers and Lakes

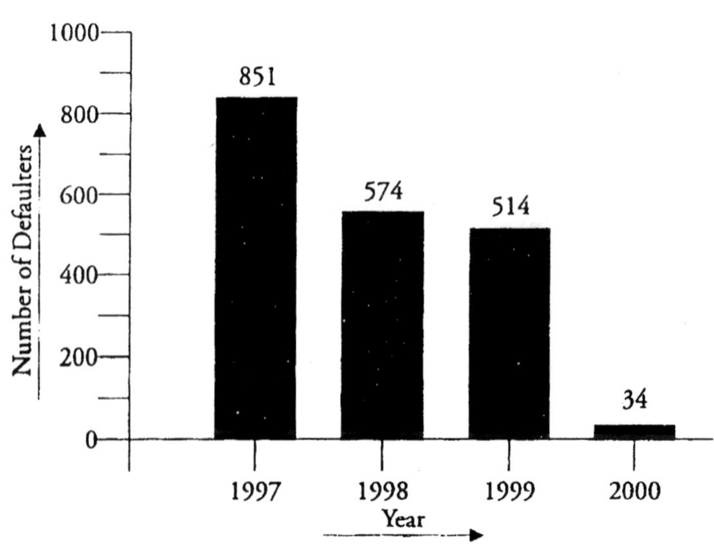

operating cost, for the implementation of all the recommendations, is high and the investment nor feasible for the industry, then these recommendations, should be implemented in phases. Priorities should be fixed and action plans with time frame should be formulated.

Follow-up Actions

Follow-up actions should be taken to check the progress of implementation of recommendations. The Environment Division of the industry should meet the other divisional heads periodically to review the progress.

THE CPCB HAS MADE A SURPRISE INSPECTION OF THE POLLUTING INDUSTRIES

Programme Description

A comprehensive programme for conducting surprise inspection of the polluting industries has been initiated in December, 1999.

Programme at the State/UT level Programme

SPCBs/PCCs were requested by CPCB, in November 1999, for constitution of the surveillance squads and carry out the surprise inspection of the polluting industries, in their respective States/UTs.

Response and visits

SPCBs/PCCs which have responded	15
SPCBs/PCCs which have constituted or already have surveillance squads	11
No. of units inspected	541

Observation

The number of visits made by the SPCBs/PCs is not very encouraging. Also, only few SPCBs/PCCs have made such visits and the SPCBs/PCCs are required to give greater attention to the surprise inspection of the polluting industries to ensure proper operation of the pollution control facilities.

Statewise Distribution of the identified 1551(large and medium) Industries of 17 Categories

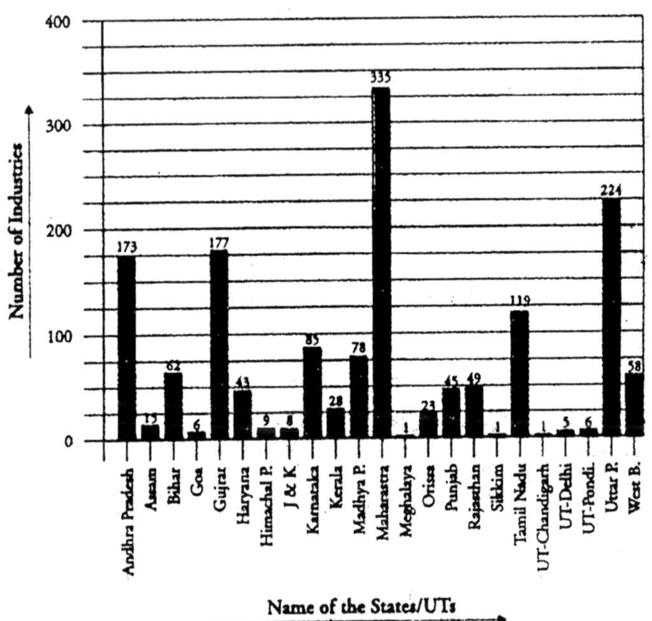

Programme at the CPCB level Programme

A total of 137 polluting industries were identified in the various States/UTs for visits by the CPCB team during December 1999 to February 2000.

Observations

Most of the industries inspected either do not have the requisite facilities or do not operate the facilities. The visit reports have been forwarded to SPCBs/PCCs for implementation of the findings and the action against the units have also been taken directly by CPCB wherever necessary. However, intensive surveillance is required at the State/UT level itself to ensure proper operation and maintenance of the pollution control systems by the industries.

Reduction in Number of defaulters among 119 units Identified along the River Ganga

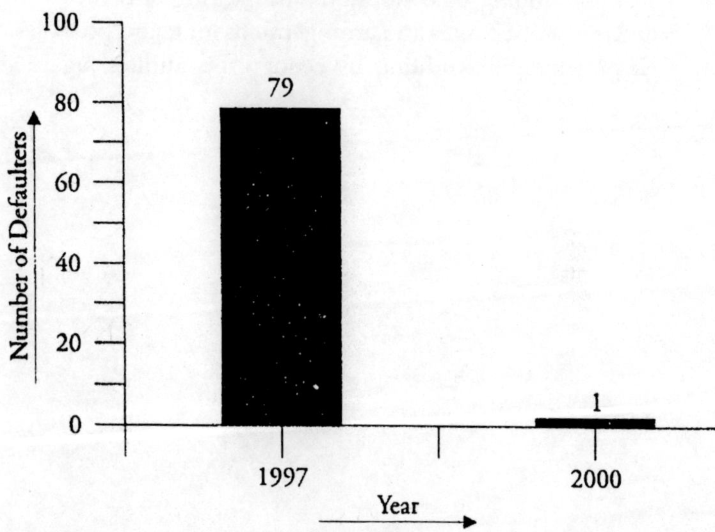

Limitations

- Long time periods involved in phasing out of the old process technologies.
- Problems of retrofitting modern control devices in old plants.
- Lack of space for construction of pollution control systems in old plants.
- Unplanned growth of Small Scale Industries.
- Lack of social responsibility in the containment of pollution.
- Pollution control still considered as wasteful expenditure by industries.
- Indiscriminate location of industries in non-conforming areas and residential areas.
- Lack of requisite enforcement machinery.

Action Required for Effective Control of Industrial Pollution

- Surprise inspection of polluting industries by surveillance squads.
- Judicious location of the industries.

- Streamlining of the consent procedures and inspection system.
- Computerisation of the consent management and inspection system.
- Commissioning of continuous monitoring and recording of emissions/discharges and arrangements for guard pond system.
- Environmental Auditing by recognised auditors/agencies.

DEVELOPMENT AND PROMOTION OF CLEANING TECHNOLOGIES

Establishment of the Indian Centre for Promotion of Cleaner Technologies (ICPC)

With the assistance of the World Bank, the Ministry is establishing the Indian Centre for Promotion of Cleaner Technologies (ICPC) as a network of export institutions which will provided the necessary input in creating a data base on available technologies their relative performance and ranking, the source (s) form where the technologies can be obtained, need investments etc. The World Bank has provided US$ 2.00 million as grant-in-aid to set up the Centre. To being with, two priority sectors namely; Industry and Energy have been identified for creating the data base. Counterpart funding to meet expenses of the staff for the initial three years period is to be met by the Government of India. A significant feature of the ICPC is that it shall provide evaluated and ranked technology options to the entrepreneurs.

Demonstration Project on Cleaner Technologies

Under the Cleaner Production Programme through Promotion of Clearner Technologies, a demonstration project on "Thermo-Chemical Conversion Reactor for utilisation of Distillery Waste" was undertaken by M/s Esvin Group of Companies. The project envisaged carrying out bench and demo plant trials before commercialisation. The trial runs have been completed and the findings submitted by M/s. Esvin to the Ministry are being scrutinised by experts. It is now proposed to commercialise the technology through a collaborative effort among the manufacturing unit(s) in the paper and distilleries sectors, Technology Development Agencies and Financial Institutions.

Adoption of Clean Technologies in Small Scale Industries

The scheme for promotion, development and adoption of clean technologies, including waste reuse or re-cycling, formulated earlier for small scale industries, is being continued. The following activities were carried out under the scheme, during the year:

- Training and awareness programmes and short duration educational workshops for small scale industry sector and for entrepreneurs in selected small scale indusries were organised through the Small Industry Service Institutes (SISIs). These programmes focussed on bleaching, dyeing and printing, electroplating, paint industry, chemical industries, lead-acid batteries, hosiery industry, plywood manufacturing, refractory industry, etc.

- Publications such as general guidelines/manuals on waste minimisation and from waste to profits have been brought out.
- Sector specific manuals on waste minimisation in respect of pulp and paper based on agriculture residues, textiles, dyeing and printing, electroplating have been brought out
- Sector specific manuals on waste minimisation in respect of pulp and paper based on agriculture residues, textiles, dyeing and printing, electroplating have been brought out
- Studies on waste minimisation and demonstration in bulk drug industry, chemical sector, textile processing units have been sponsored in Calcutta.

Waste Minimisation

A project on 'Waste Minimisation in Small Scale Industries' has been launched with the National Productivity Council as the nodal agency, under the World Bank assisted Industrial Pollution Prevention Project. The objective of the Project is to increase the awareness, interests and the confidence of Indian Small and Medium Scale Industries to take up waste minimisation activities on a sustained basis. It also aims at generating enough indigenous capacity to create an environment of self help in waste minimisation in the industries. The proposal has two components to achieve these two objectives:

- Development of a communication strategy for launching an awareness campaign on waste minimisation; and
- Establishment and running of Waste Minimisation Circles (WMC) in small scale industries in the Country.

Fifteen Waste Minimisation Circles have been established at different places of the country with the objective of providing opportunities for sharing the views and knowledge of waste minimisation/pollution prevention among the workers in clusters of industries producing the same products. Three hundred waste minimisation measures have been identified by the members of the WMCs which on adoption have resulted in reduction of pollution load to the extent of 20 to 40 per cent. In addition, cost savings to the extent of Rs.150 lakhs per annum is also expected.

Environmental Statement (as part of Environmental Audit)

A Gazette Notification on Environmental Audit issued by the Ministry in 1992, under the Environment (Protection) Act, 1986 has bade it mandatory for all the industrial units to submit an environmental statement to the concerned State Pollution Control Boards while seeking consent to operate under the relevant environmental norms. The Environmental Statements enable the units to take a comprehensive look at the industrial operations, facilitate the understanding of material flows and help them to focus on areas where waste reduction and consequently savings in material cost, is possible. Activities undertaken during the year are as follows :

- Sector wise environmental audit training programmes in respect of fertilizers, oil refinery, pesticides and food processing industry were organised through the FICCI.
- Sector-specific environmental audit manuals in respect of sugar and tube oil re-refining industry have been prepared.
- Training programmes for officials of Central and State Pollution Control Boards to make them acquaint themselves with the use of the software 'PARYAVARAN' being organised through the Centre for Environmental Studies, Anna Malai University, Chennai. Based on the inputs received during the training programmes, the software has been modified to meet the requirements of the Pollution Control Boards. The software has been distributed to all the Pollution Control Boards.
- In order to make environmental statements more effective. Forum V of the Statement has been revised for notification under the Environmental Audit Rules of the Environment (Protection) Act, 1986

Development of Environmental Standards

The Ministry has laid down industry specific as well as general effluent and emission standards under the Environment

(Protection) Act, 1986. During the year standards for Coke Plants (hydroproduct recovery type) were finalised and notified in the official Gazette. Emission/effluent standards for the following categories of industries have also been finalised:

- Battery Manufacturing Industries;
- Gas Based Power Stations;
- Standards for DG sets
- Temperature Limits for discharge of condenser cooling water from Thermal Power Plants; and
- Time limit for changeover from moving kilns to fixed chimney RCC kilns.

The compliance of these standards by industries/plants is ensured by the State PCBs/PCCs mainly through their consent mechanism.

Recognition of Environmental Laboratories under the Environment (Protection) Act, 1986 Under Section 12 and 13 of the Environment (Protection) Act, 1986 the Ministry recognises environmental laboratories and Government Analysts working in the laboratories, to carry out functions entrusted to them under the Act. While the powers for recognising laboratories of the Government and autonomous organisations have been delegated to the CPCB, laboratories in the private sector are given recognition by the Ministry. During the year, based on the examination and recommendation of the State Pollution Control Boards, two private laboratories have been recognised.

Industrial Pollution Complaints

During the year, around 680 complaints regarding various types of pollution-air, water, noise and soil have been received and attended to. Most of the complaints relate to pollution from smallscale industrial units operating either amidst thickly populated residential areas or in the immediate vicinity of human settlements. The other category of complaints relates to the discharge of untreated or partially treated effluent and emissions from large and medium scale industries which contaminate the surface/ ground water and create water logging, increases the levels of SPM and sulphur-dioxide which in turn may affect the crops and human health. Complaints have also been received about waste dumps, burning of animal carcass, operation of stone crushers, brick kilns, portable electric generator sets and air and noise pollution due to heavy vehicular traffic.

Control of Vehicular Pollution

The Ministry of Environment and Forests has recommended the notification of emission norms for passenger cars fitted with catalytic converters to be 50% lower than cars without it, meeting the 1.4.1996 emission standards, in the following manner :

Pollutant	*Emission Standard for Petrol Driven 4-wheelers*	
	Without Cat. Converter	*With Cat. Converter*
CO	8.68 - 12.40	4.34 - 6.20
HC + NOx	3.0 - 4.36	1.5 - 2.18
Total (CO + HC + NOx)	11.68 - 16.76	5.84 - 8.38

These norms have been notified by the Ministry of Surface Transport on 2.1.1998 Phased Tightening of Exhaust Emission Standards for Indian Automobiles

Category	*1991*	*1.4.1996*	*1.4.2000*
Petrol Vehicles : (in grams/km)			
Two Wheelers:			
Total (CO+HC+NOx)	20-42	8.1	4.0
Three Wheelers:			
Total (CO+HC+NOx)	20-42	12.15	6.0
Passenger Cars:			
Total(CO+HC+NOx)	16.3-30.0	11.68-16.76	3.69
Diesel Vehicles: (in gms/KWh)			
A: Gross Vehicles Weight> 3.5 ton (Heavy duty vehicles)			
Total (CO+HC+NOx+PM)	35.5	28.0	13.96
B: Gross Vehicles Weight<3.5 ton (Light duty vehicles)			
Total (CO+HC+NOx+PM)	35.5	28.0	13.96-14.57

CO : Carbon monoxide HC : Hydrocarbons, NOx : Oxides of Nitrogen, PM : Particulate Matter

Noise Pollution

Noise Pollution is a direct manifestation of increasing industrialisation, urbanisation, vehicular traffic and other commercial activities. It is as much an environmental issue as water, air and soil and pollution. Sound levels have increased over the time to the extent that it seems to be irreversible. The Central Pollution Control Board has been regularly conducting noise pollution surveys in major Indian cities. During the current year such surveys have been conducted in the cities of Varanasi and Dehradun. The data collected for Varanasi indicates that noise levels are well within the limits in all the areas except in the commercial areas where it exceeds the prescribed standards during certain times of the day. Similarly, monitoring of noise pollution for Dehradun reveals that noise levels are within the limits in the industrial and residential areas. However it exceed the standards in commercial areas and silence zones during certain times of the day.

A high level Committee on Noise Pollution has been constituted at the National Level to look into the growing menace of noise pollution in the major metropolitan cities of the country. The committee has recommended that noise levels from the bursting crackers should not exceed 100 dBA at the distance of four meters. Accordingly, directions have been issued to all the concerned authorities at the state level to check the noise pollution from bursting of crackers.

Central Pollution Control Board

The Central Pollution Control Board (CPCB) is an autonomous body of the Ministry set up in Sept. 1974, under the provisions of the Water (Prevention and Control of Pollution) Act, 1974.

It coordinates the activities of the State Pollution Control Boards (CPCBs) and the Pollution Control Committees (PCCs). The CPCB, SPCBs and the PCCs are responsible for implementing the legislation relating to pollution and control of pollution; they also develop rules and regulations which describe the standards for emissions and effluents of air and water pollutants and noise level. The CPCB advises the Central Government on all matters concerning the prevention and control of air, water and noise pollution and provides technical services to the Ministry for implementing the provisions of the Environment (Protection) Act, 1986. Details of some of the major activities of the CPCB.

CASE STUDIES

CASE STUDY-1

Organo-Chlorine Pesticides Industry (Company 'A')

The industry is located in the Northern part of the country, and is involved in the manufacture of dichloro dipheny trichlorothane (DDT), an organochlorine pesticide. It is a large-scale industry and is in operation for about 40 years. The industry was commenced when there was not so much

consciousness about pollution control. Over the years, residential development blossomed around the factory and a situation has arisen wherein the industry has to immediately review its activities thoroughly and effectively manage the wastes generated for its existence.

Environmental auditing was conducted by a team of scientists/ engineers from the Central Pollution Control Board (Central Board), Delhi. This was the first pesticides industry to be audited by the team. The procedure followed and the results obtained are explained below.

PRE-AUDIT ACTIVITIES

Preliminary information

A questionnaire was sent to the industry and preliminary information obtained on products manufactured, raw materials and water used, wastewater, gaseous emissions, and solid wastes generated, waste disposal points and details on surrounding land uses. Details in brief based on the information received are given below.

The industry is involved in the manufacture of technical grade DDT and its formulations. The raw materials used for technical grade DDT are benzene, alcohol, chlorine and oleum

(20%), and those used for formulations are : technical grade DDT, china clay and soap stone. The manufacture of DDT involves broadly the following three steps :

Step I : Manufacture of monochlorobenzene (MCB)
Step II : Manufacture of chloral
Step III: Manufacture of DDT

Step I has the unit operations of drying of benzene, chlorination of benzene, neutralisation and washing of chlorobenzene, enrichment of chlorobenzene and distillation of monochlorobenzene (MCB).

Step II has the unit operations of chlorination of alcohol, distillation of chloral alcoholide in the presence of oleum, and absorption of hydrogen chloride.

Step III includes the unit operations of condensation reaction between MCB and chloral using oleum as a condensing agent, separation of aqueous and organic phases, recovery of MCB from organic phase, and solidification of DDT.

The flow sheet of various operations as provided by the industry is shown here with. The total water consumption is 460 kld of which 270 kld is for process, 30 kld for sanitary, 10 kld for services, 125 kld for boiler and 25 kld is for cooling water make-up.

The wastewater generated is 290 kld of which 250 kld is from process, 30 kld from sanitary and 10 kld from boiler blow down. Steam is borrowed from a neighbouring industry and hence boilers are not in regular use. The wastewater is characterised by parameters like, pH, SS, COD, BOD, O&G, sulphates, chlorides and DDT. Solid waste generated is 18 t/month (70% water) from effluent treatment plant (ETP) and is disposed off into a municipal dumping yard.

The gaseous emissions include combustion emissions from usage of furnace oil at the boilers whenever required and process emissions of benzene, chlorine, HCl and dust of DDT. The industry reported its problems to bring down COD within permissible limit of 250 mg/1 and disposal of sludge from ETP.

Audit team

An audit team of seven members was constituted. The team included a Senior Environmental Engineer, two specialists on pesticides industry-one scientist and the other an engineer, two members for water sampling, and two members for emissions sampling. The members with relevant experience were chosen based on the nature of work assessed from the preliminary information. The members had expertise in their own fields. The two specialists were aware of environmental problems associated with a pesticides industry and various waste management options. From the factory side, an overall coordinator and representatives from production, environment management and laboratory were identified to assist the audit team. The members were allocated with specific tasks.

Resources

The laboratory of the Central Board at Delhi was chosen for analysis of samples. This laboratory has the facility to analyze various parameters identified, including DDT. The instruments required for sampling were high volume sampler and handy sampler for ambient air quality and stack monitoring kit for emissions form stacks .These were made available by the above laboratory.

Visit programme

The industry being a single product plant having no complications as in multi product plants, a visit programme for 3 days was prepared. The programme was communicated to the industry well in advance, and also informed to confirm back that various operations would be normal and that there would be no shut downs or partial operations during the auditing period. After receiving confirmation, the visit programme was communicated to the members and the laboratory.

Activities at the site

On reaching the site, a meeting with the management of the industry was held. The purpose of audit was explained to them. The coordinator of the industry introduced various representatives who would assist the audit team. The manufacturing processes and various activities that take place at the site were explained to the audit team. The team then made a reconnaissance survey of the industry so as to be acquainted with location of various activities and operations. The team then went ahead with other activities at the site which are a part of the audit.

Material balance

The chemical reactions involved in the process are given below :

C_6h_6 + C12 ⟶ C6H5C1 + CL
(78.1) (70.9) (112.5) (36.5)

2 C_2H_5OH + 4C12 ⟶ CC13CH(OH) OC_2H_5 + 5HCL
(92.1) (283.6) (193.5) (182.2)

oleum

CC13CH(OH) OC_2H_5 ⟶ CC13CHO + C_2H_5O SO_3 H
(193.5) (147.5) (126)

$H_2SO_4.SO_3$

CC13CHO + 2 C_6H_5CI ⟶ $(C_6H_5Cl)_2$ CHCC13 + H_2O
(147.5) (225) (354.5) DDT (18)

The process flow chart incorporating various unit operations alongwith inputs of raw material and water and outputs of products, by-products, wastewater, gaseous emission, solid waste and intermediates is shown in here with. The quantities of inputs and outputs at each unit operation could not be worked, as the industry had no ready data. As it is a long procedure and time consuming, the industry was explained how to carry out the work, and was asked to complete the work later. The mass balance of chlorine and oleum derived from the overall usage is given below.

The requirements of raw materials per tonne of products are given as below :

RAW MATERIAL REQUIREMENT – STOICHIOMETRIC VS. ACTUAL

Raw material	*StoichiometricActual Requirement (kg.)*	*Requirement* (kg.) consumption*	*% Extra*
Benzene	440.7	780.	77.1
Chlorine	1200.2	1773.24	47.7
Ethanol	259.9	409.00	57.4
Oleum (20%)	-	1580.58	-

(* Based on the average of four years)

The total water requirement is 460 kld and the wastewater generation is 290 kld. The water balance sheet is given in Fig 4.2.

Waste flow

The wastewater generated form process is 30 kld form HCl scrubber of MCB plant , 30 kld from HCl scrubber of chloral (CA) Plant, 140 kld from washing and neutralisation of DDT condensation, 40 kld form spent sulphuric acid recovery plant and 10 kld form floor washing in various plants. Pretreatment units include a separator in MCB plant to remove floating organic a DDT separator to settle down DDT form the wash stream (140 kld) and separate settling tanks for wastewater from CA and MCB plants and DDT condensation plant . The wastewater then flows though a final effluent treatment plant (ETP) comprising equalization tank, neutralisation with lime, clarifier and sludge drying beds The wastewater is then discharged to a public drain.

The domestic wastewater is passed through septic tanks and disposed into the public drain. The wastewater flow lines are shown here with.

The gaseous emissions include benzene, chlorine, HCl, alcohol vapours, DDT and fugitive emissions from process. The source of emissions, its type and details of control equipment provided are given here with.

The main source of solid waste is ETP. The quantity of waste generated is 18 t/d on wet basis containing 70% water. It is disposed to the municipal dumping yard. Raw material carrier gunny bags are burnt in the factory and drums are sold out.

Monitoring & analysis

Sampling points identified for analysing wastewater and performance of various treatment units are shown here with 4-hr composite sampling was made. Parameters analysed to identify the characteristics of wastewater and to evaluate performance of various treatment units and the analysis results are incorporated here with.

The stacks could not be monitored as they were not installed properly and stack monitoring facilities viz. Porthole etc. was not provided.

PROCESS EMISSIONS - TYPE, SOURCE AND CONTROL EQUIPMENT

Source of Emission	*Type of Emission*	*Control Equipment Provided*	*Remarks*
MCB Section Chlorinator	• Benzene Vapours	i) Benzene scrubber	No provision in the chimney for monitoring. Emissions to be monitored and suitable chimney height to be provided
	• Chlorine Gas	ii) Water absorber to scrub HC1 and produce 20% acid as by-product	- do -
	• HCl gas	iii) Tail gas absorber where HC1 gas is further scrubbed with water	- do -
CA House Chlorinator	• Alcohol Vapours	i) Alcohol vapours removed in a condenser using cooling water and brine	- do -
	• Chlorine Gas	ii) Water scrubber to absorb HC1 gas and produce 30-35% acid as by-product	- do -
	• Chlorine Gas	ii) Water absorber to scrub HC1 and produce 20% acid as by-product	- do -
	• HCl Gas	ii) Tail gas absorber where HCl gas is further scrubbed with water	- do -
Grinding Section	• Dust of DDT and Inerts	Dust extraction systemWith hoods, ducts, cyclone separator and a chimney of 55 m height	- do -

The ambient air quality was monitored at seven locations within the factory premises including the process areas using high volume samplers and handy samples. The results are given here with.

The public drain where the treated wastewater is discharged was monitored at upstream and downstream of the disposal point. Also, the ground water in the factory premises was monitored. The results are given here with.

The solid waste was analysed for DDT and DDE (dichloro diphenyl ethylene) - a breakdown product of DDT. The analysis of this waste showed DDE (PP') of 15474 mg / g, but DDT could not be traced.

Field observations

- In the manufacturing process certain side reactions take place. During the chlorination of benzene, dichlorobenzene (DCB) is also formed alongwith MCB which is recovered and sold as a by-product to the deodorant products manufacturers. During the condensation process of chloral and MCB to form DDT, a small amount of chlorobenzene sulphuric acid is also formed. These side reactions lead to wastage of materials. H_2SO_4 is used as a catalyst and for absorbing water. It is the major pollutant in the wastewater as the spent sulphuric acid is not recovered entirely. Certain process modifications are deemed necessary.
- The wastewater collection and drainage system is poor. No separate storm water drainage system has been provided. The process wastewater drains are subject to entry of stormwater due to which handling of wastewater during rains becomes very difficult.
- The manual dosing of lime in the neutralisation tank of ETP is inefficient. The operators though are able to test pH, adjusting of lime dosage is not well known.
- The sludge drying beds of ETP are choked. The solid waste is not properly collected. Scrap was found lying scattered in the plant. There is no earmarked area for solid waste.

Ambient Air Quality

Sl.	Location	Sampling Period	Total Sampling Time (minutes)	Analysis Results(mg/m^3) SO_2	Cl_2	SPM
1.	Near chloral Plant	11.55 AM - 03.44 PM	240	61.8	8.0	306
2.	Near DGM building at the height of 9 m.	12.10 PM - 14.10 PM	240	43.5	2.7	276
3.	Near GMBuilding at theHeight of 4 m.	1.10 PM - 05.10 PM	240	46.3	3.6	232
4.	Near Mono-chloro benzene(MCB) unit	11.40 AM - 01.40 PM	120	-	12.5	-
5.	- do -	01.45 PM - 03.45 PM	120	-	ND	-
6.	Near duct tower	11.40 AM - 01.40 PM	120	-	16.1	-
7.	- do -	01.45 PM - 03.45 PM	120	-	3.6	-

Groundwater Analysis

(i)	pH	:	6.3
(ii)	Chlorides (mg/l)	:	2101
(iii)	Sulphate (mg/l)	:	573
(iv)	Total Hardness (mg/l)	:	3480
(v)	Conductivity (mmhos/cm)	:	7380
(vi)	Phosphate (P) (mg/l)	:	Not traceable
(vii)	Nitrate (N) mg/l)	:	0.46
(viii)	Alkalinity (mg/l)	:	463
(ix)	DO (mg/l)	:	Not traceable
(x)	DDT PP' (mg/1)	:	14569
(xi)	DDT OP' (mg/1)	:	3627
(xii)	DDE PP' (mg/1)	:	2829

- Chemical spills overflows and leaks from pipes in the process areas were observed almost everywhere. Final product of DDT ,due to a crack in the DDT casting pans, was finding its way to wastewater drain.
- The HCl gases are scrubbed with water and discharged to ETP, This wastewater and that form washing in condensation reaction of DDT is contributing to acidity due to which huge quantities of lime have to be used for treatment. The source of DDT in the effluent is from washing in condensation section of DDT(140kld).
- Huge quantities of HCl fume were observed while loading HCl tanker. Also emissions were observed from the HCl storage tanks.
- Housekeeping is poor. Proper colour codes, sign boards and instructions are not incorporated. The industry has no clear-cut buffer zone or green belt with the surroundings.
- Residential areas surround industry. Land availability is limited at the site. Site layout is poor.
- Industry has of late taken steps to improve waste management. However, the industry staff are of their opinion that since it is a very old factory it is very difficult to make modifications needed for the best practicable waste management. Industry has a separate R&D wing to carry out research for improvement. The staffs are not aware of the latest environmental management techniques and waste management options.

Draft Report

No draft report was prepared at the site as the audit guidelines were yet to be evolved as this industry was the first industry to be audited. However, detailed discussions were carried out with the management about the observations and possible recommendations. The management agreed with the viewpoints of the audit team and welcomed a detailed report with recommendations for implementation.

The comparison of the stoichiometric and actual requirements based on average figures of four consecutive years shows that there is excess usage of chemicals like benzene - 77% chlorine-48% and ethanol - 57%. About 258kg/d of HCI (100%) and 3.7 t/d of H_2SO_4 (100%) are finding their way to environment. The excess usage of materials besides contributing to loss of economy are causing pollution. The excess usage may be attributed to inefficient process technology poor process performance poor house keeping, lack of implant control measures and poor material handling

The wastewater generated is about 40 kld per tonne of product manufactured. It contains sulphuric acid, hydrochloric acid, soluble organics in the form of benzene, alcohol, MCB, DCB and DDT. The effluent is highly acidic and toxic in nature. The major contributor to water pollution is soda and hot water wash in DDT condensation section. From DDT cast pans, due to leaks, an amount of 6.6 kg/d of DDT is being contributed to ETP. HCl in the effluent is contributed form HCI scrubbers and H_2SO_4 is contributed from Soda and hot water wash and H_2SO_4 recovery plant.

The solid waste generated is hazardous in nature due to the presence of DDE and hence cannot be disposed to municipal dumping yard. Also an authorisation form the pollution control agency is to be obtained for the disposal of this waste. The open burning of used raw material bags is not permissible as it causes air pollution and may generate toxic fumes.

The scrubbers provided for HCl emissions appear to be insufficient. Additional caustic soda scrubbers are to be provided. The fugitive emissions due to material handling, handing from storage tanks and form loading and unloading of tanks are to be controlled.

The wastewater treatment system provided is insufficient. The DDT separator is not serving any purpose mainly due to insufficient detention time provided for DDT to settle down. The manual dosing of lime in neutralisation tank of ETP is inefficient. The clarifier is under- designed. The treated wastewater is not conforming to the prescribed standards.

DDT concentration in the wastewater can be reduced by 50% by controlling leak form the DDT cast pans and collecting overflows or spills occurring while sampling under DDT Condensation vessel.

- DDT separator is to be modified to achieve proper detention time for settling. Alum Coagulation aided settling may enhance rate of DDT.
- The groundwater in the factory premises is highly polluted. It has a DDT of about 20,000 mg/1 and also high concentration of conductivity and chlorides. The process wastewater and leachate form sludge dumping might have caused this pollution of groundwater.
- Industry has air pollution problem due to HCl fumes from process. HCI storage tanks and loading of HCI road tankers. Fugitive emissions were observed at various process sections. On - line gases pose air pollution problem during power failures since stand-by D.G. set is not available.
- House-keeping is poor. Spills. leaks and fumes are generated from various processes, Drainage system and collection of scrap and waste are poor.

Final report

A final reports incorporation various observations and recommendations were prepared. The important recommendations include the following.

- Review of process technology adopted and performance of process equipment is deemed necessary to take up modifications accordingly. Fixing up of norms for performance of process operation and wastes generated imminent such that the loss of materials and hence the wastage are minimised. The possibilities of replacing oleum with a suitable alternative may be explored.

- The proposed scheme for waste treatment is shown here with.
- The top level of process drains are to kept atleast 15 cm above ground level so that the rain water does not enter these drains. Storm water drainage is to be kept separate and polluted storm water carrying spills during the initial hours of rain fall to be collected and treated. Flooring in the process areas shall be made properly. Treated wastewater to be recycled for floor wash, making lime solution and for gardening purposes.
- Tree plantation in the surrounding shall be developed to act a buffer zone and also to improve the environment.

CASE STUDY-2

Organophosphorus Pesticides Industry (Company 'B')

Company 'B' is a large scale industry involved in the manufacture of technical grade organophosphorus pesticides and their formulations. It is relatively a new industry and hence environmental aspects were given due consideration at the design stage . Auditing was conducted in this industry to probe how best various systems were working and to look into possibilities for further improvement.

PRE-AUDIT ACTIVITIES

Preliminary information

A questionnaire was sent to the industry for collection of preliminary information. Details based on the information received are given below;

Industry is involved in the manufacture of basic pesticides like monocrotophos cypermethin and dichlorouos and their formulations. It is located in an industrial area. The steps involved in the manufacture of above product are as follows;

(a) Monocrotophos (MCP)

Step I: Dehydration of monomethyl acetoacetamide (MMD) to obtain anhydrous MMA

Step II: Reaction of MMA with chloral in a toluene medium to obtain MMA adducts.

Step III: Filtration and dissociation of chloro adduct to form MMACL

Step IV: MMACI is reacted with trimethyl phosphite (TMP) in dichloroethane medium to Form monocrotophos. Methyl chloride gas is evolved.

(b) Cybermethrin (CYPER)

The first step involves condensation of 2-dimethyl –3 (2-2 dichlorovinyl) –cyclopropane-1, carboxyl chloride (DVO), Meta phenoxy benzaldehyde (MPB) and sodium cyanide in xylem media in the presence of a catalyst. The above mixture is washed to remove ionic impurities. The washed product is dried and purified by distillation. Solvent is recovered and recycled. The product is cooled and filtered to obtain cypermethrin.

(c) Dichlorovos (DDVP)

Chloral and TMP are reacted to produce crude DDVP that is further distilled to obtain required purity.

(d) Phosphamidon (PMN)

Step I : Diethyl acetoacetamide is reacted with chlorine to produce diethyl dichloro- acetoacetamide (DDA)

Step II : DDA is reacted with TMP in presence of chlorobenzene to crude amide phosphamidon

Step III : Crude phosphamidon is vacuum distilled to preconcentrate and is further flashed in a thin Film evaporator to produce phosphamidon of required purity.

The water requirement is 1160 kld of which 140 kld is for boiler, 220 kld for cooling water, 160 kld for process (92 kld for monocrotophos, 60 kld for phosphamidon/DDVP and 8 kld for cypermethin), 200 kld for sanitary and 440 kld for services (fire, gardening, etc.). The wastewater generation is 500 kld for which 140 kld is from boiler blow down, 200 kld from sanitary and 160 kld is from boiler blow down, 200 kld from sanitary and 160 kld from process (MCP: 92 kld, PMN/DDVP: 60 kld, CYPER : 8 kld). The wastewater characterised by parameters like pH, BOD, COD, TSS, cyanides, residual chlorine, pesticides and toxicity. The wastewater treatment includes pretreatment of process wastewater for detoxification followed by combined treatment of domestic and process wastewater in a two-stage biological treatment plant.

The process emissions include HCl, CH_3Cl, chloroform, HCN, P_2O_5 and SPM. Gaseous emissions also arise from combustion of furnace oil at boilers.

The solid waste generated is about 5.0 t/yr. It contains oils, pesticides, toxic residues, discarded containers etc.

Audit team

The audit team had seven members, same as the earlier Case Study.

Resources

The industry has a full-fledged laboratory to analyze various expected parameters, including BOD, Pesticides and Toxicity. The laboratory also has stack monitoring kit and high volume samplers. The laboratory of the Central Pollution Control Board at Delhi was chosen as a standby.

Visit Programme

A programme for a visit to the industry was prepared for six days. The programme was communicated to the industry well in advance and got confirmation of the normal working during those six days.

Activities at the site

A meeting with the Management of the industry and the representatives from the industry who would assist the team was held and a reconnaissance survey of the industry, was done. The team then went ahead with performing various other activities as detailed below.

Material balance

The chemical reactions involved in the manufacturing process are as follows :

The process flow charts for the above products incorporating various unit operations alongwith various inputs and outputs of products, wastewater, gaseous emissions, and solid waste are given here with. The requirement of raw materials per tonne of each product and the details of wastes generated alongwith pretreatment details are given in Tables and the total water requirement is 1081 kld and 570 kld is the wastewater generation. The water balance is shown here with.

Waste flow

The sources of the wastewater are identified and their quantities determined. The wastewater from monicrotophos is 160 kld, kld from cypermethrin, 60 kld from phosphamidon and 16 kld from DDVP. The other wastewaters are 50 kld from incinerator, 28 kld from boiler blow down, 175 kld

a) **Monocrotpphos:**

```
                H                                                 H
                |                                                 |
i)    CH3— CO – C – CO – NHCH3 + CCl3 CHO ---------------> CH3-CO-C-CO-NHCH3
                |                                                 |
                H                                              HOHCCl3

             (MMA)               (Chloral)                   (MMA Abduct)

ii)   MMA Abduct  +  Cl2 -----------------> CH3 – CO – C – CO – NHCH3 + HCl
                                                       |
                                                    HOHCCl3
                                                 (Chloro abduct)

                                             Cl
                                             |
iii)  Chloro abduct ---------------> CH3-CO-C-CO-NHCH3 + chloral
                                              |
                                              H
                                           (MMACl)

                    OCH3                    O    H
                     |                      |    |
iv)   MMACl + OCH3 – P – OCH3 -------> OCH3-P-O-C =C-CO-NHCH3 + CH3Cl
                                            |    |
                    (TMP)                   OCH3 CH3
```

b) **Cypermethrin :**

```
                                                    Catalyst
Acid Chloride + m-Phenoxybenzaldehyde + NaCN ----------->

               NaCl        + Cypermethrin
```

c) **DDVP:**

```
                                                  O
                                                  |
CCl3CHO  + (CH3O) 3P ----------> (CH3O) 2P-O-CH  = CCl2 + CH3Cl
```

Table : Data Sheet on inputs and outputs for the Manufacture of Cypermethrin

Raw material per tone of product

Sen.	*Raw material*	*Quantity, kg* *Actual*	*Stoichiometric*
1.	Acid Chloride (DVO)	580	547
2.	Metaphenoxy benzaldehide	490	476
3.	Sodium cyanide	208	118
4.	O-xylene	960.5	Not process Chemicals used for washing
5.	Na_2CO_3	161.5	
6.	NaCl	280	

Water requirement per tonne of product

Steam : Nil (no jet-ejectors)
Cooling : Nil
Process : Washing with Na_2CO_3/NaCl : 675 l
Making solution of NaCN : 2501

S.No.	Source	Nature	Quantity, Kg	Pretreatment
1.	Washing with Na2CO3/NaCl	NaCN, Cypermethrin NA2CO3, NaCl	675	Detoxification Using H202 a maintaining p using acetic a
2.	Caustic Scrubber for HCN emission	Scrubber liquor	Varying Quantity	- do -
3.	Drips and Drains	-	- do-	- do-

S.No.	Source	Nature	Quantity	Gas flow Nm3/hr	Control equip.	Chimney Ht., m.
1	Conden sation	HCN	-	900	Caustic Scrubber	15

Solid Waste/hazardous waste: Nil

Data Sheet on inputs and outputs for the Manufacture of Dichlorovos

Raw material requirement per tonne of product :

S.No.	Raw Material	Quantity, kg Actual	Quantity, kg Stoichiometric
1.	Chloral	771	668
2.	Trimethyl phosphite	570	561

Water requirement per tonne of product :

Steam : 4.0
Cooling : 5.0
Process : Nil
Total : 9.0 kl

Waste Water generated per tonne of product :

S. No.	Source	Nature	Quantity	Pretreatment
1.	Jet condensate & spills	Solvent traces of toxic material	4.0	Detoxification with NaOH at pH 10-10.5

Emission:

S. No.	Source	Nature	Quantity	Gas flow	Control equip.	Chimney ht., m
1.	Chlorination	HCl	244	3500	Water scrubber	39
2.	Toxification	CH_3Cl	168	17760	Nil	-do-

Hazardous waste/Solid waste : Nil

Hazardous waste:

S. No.	Source	Nature	Quantity, kg	Pretreatment	Method of disposal
1.	Preconcentration	Degraded HC	216	Nil	Incine-ration
2.	Product Purification				

from laundry and toilets and 37 kld from drips, drain, drum washings, etc. Raising pH to 10-12 using caustic soda, and sufficiently detaining the wastewater under agitation detoxifies the toxic streams. The wastewater from various sources is then collected and further treated in effluent treatment plant (ETP). The wastewater flow lines are given in Table.

The process emissions include HCl, CH_3Cl, chloroform, CH3Cl, HCN, P_2O_5, SPM, SO_2 and Nox. The details of the sources of emissions, gas, flow rate, control equipment provided and height of the chimney are given in Table. About 5.0 ft of solid waste is generated per annum. The details of its type, quantity and disposal are given in Table.

Monitoring and analysis

The sampling points identified to determine the characteristics of wastewater and performance of wastewater treatment system are shown here with. The stacks could not be monitored due to time constraint. Two ambient air quality stations are operated by the industry at the site. The available data with the industry was collected. The process samples were analysed for pH, SS, COD, BOD, TDS, chlorides, TOC, TSS, pesticides (MCP, CYPER, DDVP, PMN) and flow rate, and DO, MLSS, and MLVSS in case of aeration tanks.

Process samples were of grab type but the ETP samples were collected hourly for eight hours and composited. The samples were collected on two consecutive days.

Field Observations

Not all the products are manufactured throughout the year. Monocrotophos is manufactured throughout the year but cypermethrin is manufactured for nine months. Phosphamidon for eight months and dichlorovos for two months. Monocrotophos and cypermethrin have separate process route but phosphamidon and DDVP have only one process route due to which only one of these products is manufactured at a time. Monocrotophos is manufactured at 1.7 t per batch and batch time is 8.00 hrs, whereas phosphamidon is manufactured at 1.95 t per batch and the batch time is 16 hrs, cypermethrin manufactured at is 1 t/batch and batch time is 2 days, and DDVP is produced at 3.45 t/batch and batch

Details of Hazardous waste

S.No.	*Type of waste*	*Quantity Kg/yr*	*Method of disposal*
1.	Cyanide waste from used cyanide	0.05-0.1	Empty container along with plastic bags is filled with H_2O_2, detoxified and then buried with alkali
2.	Organic residues	165 kl/yr	Incineration
3.	Oil drips as drains	200-300	Collected using saw dust, and incinerated
4.	Sludge from ETP	3600	Land filling in factory premises
5.	Waste containing pesticides	195 kl (10.5-3% of pesticides)	Incineration
6.	Off-specification and discarded products		Incineration
7.	Discarded containers and container liners of hazardous and toxic waste	1200	Metalic drums are detoxified by heating at high temperature and then crushed and sent as scrap

time is 16 hrs. Cypermethrin manufactured at is 1 t/batch and batch time is 2 days, and DDVP is produced at 3.45 t/batch and batch time is 16 hrs.

- These variations in daily production lead to variation in characteristics of wastewater Received at ETP. Detoxification systems are provided at the process wastewaters to Reduce shock loads due to toxicity on the ETP.
- The wastewater from various sources is collected through a well-designed drainage System. Stormwater drains are kept separate.
- The treated wastewater is disposed to a public sewer, which finally joins a river. The river has a back waster normally except during rains.
- The location where the industry is existing has not been declared as air pollution control area Under the Air Act, 1981 and hence this Act is not applicable. Industry need not obtain consent for emissions from the State Pollution Control Board.
- Industry has good house-keeping practices. Safety and quality control are given highest priority. Floor washing is totally avoided. In case of spills, saw dust is used to wipe it out. Overflows, spills and leaks are kept to the minimum. Manual transfer of chemicals is avoided. Site is well laid out considering compatibility and safety while locating various activities.
- The site is surrounded by hills and valleys.
- The staffs are qualified; process performance norms are well understood. However, the Operators at the ETP are not fully aware of 'don'ts
- Land is available in plenty for any addition, if necessary.

Draft Report

A draft report was prepared and handed over to the management. The management was not in favour of recovering a contaminated by product and reusing or recycling waste that can contaminate product, but welcomed suggestions for the improvement of environment and assured immediate implementation.

POST-AUDIT ACTIVITIES

Synthetics

(i) A comparison of stoichiometric and actual requirement of various chemicals shows that their excess usages are 32-55% in case of monocrotophos. Sodium cyanide use is very high to the order of 76% in excess in case of cypermethrin, and trimethyl phosphite used is about 56% excess in case of phosphamidon. The material usage is not much in excess in case of dichlorovos. These excess usages of materials may be presumed to be finding their ways to environment thereby causing pollution. The cause of these excesses may be attributed to inefficient process performance and raw materials impurities.

(ii) The water consumption per tonne of product is highest with 81.4 kld in case of monocrotophos, 50 kld in case of phosphamidon, 9 kld in case of DDVP and 1 kld in case of cypermethrin. The wastewater generation from process is 56 kld per tonne of monocrotophos, about 1 kld per tonne of cypermethrin, 4 kld per tonne of DDVP and 35 kld per tonne of phosphamidon. The total wastewater generation is 540 kld of which 48% is from process, 31% is from laundry & toilets and 21% from boiler blowdown and steam condensate. The process wastewater is biodegradable but toxic. The treated wastewater is conforming to the prescribed standards except for COD which is slightly exceeding the limit of 250 mg/l. The wastewater flow rate in the final outlet is seen to be widely varying. The operators are not aware of how the flow should be regulated in the ETP.

(iii) The treated effluent is disposed to a public drain which ultimately joins a river.

(iv) About 3 t/d of methyl chloride emission, which is highly toxic is let into the atmosphere, thereby adding to the risk of environmental hazard.

(v) The ambient air quality record maintained by the industry does not include all the air pollutants emitted.

Final report

A final report incorporating various observations and recommendations was prepared. The important recommendations included the following :

- The excess usage of raw materials especially in case of monocrotophos, sodium cyanide in case of cypermethrin and trimethyl phosphite in case of phosphamidon are to be reduced. Performance studies of various process equipment are to be carried out and norms fixed up for each operation such that maximum utilisation of materials takes place and only the unavoidable wastes are generated.
- The de-aerator overflow of the steam condensate has been found to be within limits except for pH. This stream may be isolated, neutralised and reused for floor wash or used for gardening or irrigation purposes in the factory premises at the rate of 35 kld/hectare/day.
- The treated wastewater from ETP should be subjected to tertiary treatment by activated carbon, ozonation etc. so as to reduce COD and pesticides, and recycled for developing green belt in the factory premises.
- Methyl chloride emission should be recovered as a by-product through a liquidation plant or incinerated.
- Rate of flow to aeration tanks in the ETP should be kept constant to avoid shock loads.
- Ambient air quality monitoring in the factory premises should include the parameters of HCl, CH_3Cl, HCN and Cl_2 in addition to SPM, CO, NO_x and SO_2.
- A manual for operation of the waste treatment facilities is to be prepared for use mainly by the operators.
- An organisational set up for environmental management is to be made including people from production, R&D, quality control/laboratory, management, safety, waste treatment facilities and an environmental specialist. The operators of the waste treatment facilities are to be well-trained.

REFERENCES

1. Industry and Environment, UNEP, Vol. 11 No. 4, Oct/Nov/Dec, 1991.
2. Audit and Reduction Manual for Industrial Emissions and Wastes, UNEP, 1991

3. Bhattacharya, R.N., Environmental Audit in Industries : Guidelines, Ministry of Environment & Forests sponsored course on Environmental Impact Assessment held in Zoological Survey of India, Calcutta, January 18 - February 19, 1993.
4. Environmental Audit at M/s NOCIL - Agrochemicals Plant, Central Pollution Control Board, Delhi 1993
5. Raghu Babu, N. Basu, D.D. Chakrabarti, S.P., Environmental Audit in Industry, Seminar on ' Environmental Statement (Audit)' held on 29.7.93 at Alwar by Rajasthan Pollution Control Board.
6. Raghu Babu, N., Chakrabarti, S.P., Environmental Audit - a tool for Waste Management, Seminar on ' Environmental Triangle - Industry, Institutions and Regulatory Agencies' held on 7.10.93 at New Delhi by Shriram Institute for Industrial Research.
7. Minimal National Standards - Pesticides Manufacturing and Formulation Industry, Central Pollution Control Board, Delhi, COINDS / 15 / 1985-86.
8. Revised Minimal National Standards for Pesticide Manufacturing and Formulation Industry, Part I, Central Pollution Control Board, Delhi, COINDS/30/1988-89

SEPARATION TECHNOLOGIES FOR REMOVAL OF ORGANIC AND PESTICIDAL CHEMICALS FROM WASTEWATER

The following unit operations have grouped under the Separation Technology for removal of organic and pesticidal chemicals form wastewater:

1. Absorption
2. Absorption including bubble adsorption
3. Centrifugation
4. Clathration
5. Coagulation
6. Coalescence
7. Condensation
8. Cyclonic Action
9. Desorption
10. Dialysis
11. Diffusion process
12. Eelectro-photesis
13. Evaporation
14. Extraction
15. Filtration
16. Flash Expansion
17. Flotation
18. Foam Fractionation
19. Gravity Settling
20. Impringement
21. Membrane Permeation
22. Precipitation
23. Reverse Osmosis
24. Scrubbing
25. Stripping
26. Ultra-filtration

DESTRUCTION & DETOXIFICATION TECHNOLOGIES FOR TOXIC WASTES

1. Chlorine dioxide oxidation
2. Dye sensitized photo-oxidation
3. Electro-chemical oxidation
4. Flameless catalytic oxidation
   (Low temperature vapour oxidation)
5. High energy radiation
6. Hydrogen peroxide oxidation
7. Incineration/ combustion process
8. Micro-biological and other metabolic systems
9. Ozonation and other ultrasonic energy
10. Photo-decomposition- ultraviolet radiation
11. Potassium permanganate oxidation
12. Pyrolysis
13. Reductive dechlorination
14. Ultraviolet ray assisted ozonation
15. Wet catalytic oxidation

WATER USAGE & WASTEWATER GENERATION IN VARIOUS INDUSTRIES

Name of Industry	Water use	Wastewater Generation
Caustic Soda		
Mercury cell process	5 cum/t of caustic soda produced (excluding cooling water) & 5 cum/t of caustic soda produced for cooling water	4 cum/t of caustic soda (Mercury bearing) 10% blow down permitted for cooling tower
Membrane Cell Process	5 cum/t of caustic soda including cooling water	1 cum/t of caustic soda excluding cooling tower blow down
Textile		
Manmade fibre		
i) Nylon & polyester	170 cum /t of fibre produced	120 cum/t of fibre produced
ii) Viscose rayon	Limits specified in rayon grade pulp & paper applicable	
Tanneries	30 cum/t of raw hide	28 cum/t of raw hide
Natural rubber	6cum/t of rubber	4cum/t of rubber
Starch, glucose & related products	10cum/t of maize crushed	8 cum/t of maize crushed

Caustic Soda		
Mercury cell process	5 cum/t of caustic soda produced (excluding cooling water) & 5 cum/t of caustic soda produced for cooling water	4 cum/t of caustic soda (Mercury bearing) 10% blow down permitted for cooling tower
Membrane Cell Process	5 cum/t of caustic soda including cooling water	1 cum/t of caustic soda excluding cooling tower blow down
Textile		
Manmade fibre		
i) Nylon & polyester	170 cum /t of fibre produced	120 cum/t of fibre produced
ii) Viscose rayon	Limits specified in rayon grade pulp & paper applicable	
Tanneries	30 cum/t of raw hide	28 cum/t of raw hide
Natural rubber	6cum/t of rubber	4cum/t of rubber
Starch, glucose & related products	10cum/t of maize crushed	8 cum/t of maize crushed

EMISSION STANDARDS FOR SOME SPECIFIC POLLUTANTS

The emission standards for some of the specific pollutants are yet not laid down by the Central Board. These pollutants generally pertain chemical process industries and are emitted not through the conventional stacks or chimney but mostly form the reaction vessels, scrubber outlets and other such equipment. The emission of these pollutants gives rise to environment pollution and nuisance especially in the vicinity of the industrial plant. It is therefore, necessary to determine and lay down the emission limits for these pollutants .

In the absence of adequate literature source which could be helpful in specifying the emission limits for these pollutants, the Threshold limit Value (TLV) has taken as the basis to determine the emission limits. The following formula has been adopted for this purpose :

$$\text{Emission limit} = \frac{\text{TLV} \times (100 \text{ to } 150)}{3 \times 10}$$

$$= 3.33 \text{ TLV to } 5.0 \text{ TLV}$$

The constants appearing in the above formula are explained as under :

* The constant 3 is included to reduce the allowable concentration to take care of 24 hours exposure. This is to extrapolate the 8-hour span used in TLV over the span of 24 hours.
* The constant 10 is included to reduce the concentration in the ambient air to impart additional safety. This is in line with the application factor commonly used in fixing water quality criteria.
* The constant 100 to 150 represent the minimum dilution available in open atmosphere

ANNEXURES

ANNEXURE-I

MINISTRY OF ENVIRONMENT AND FORESTS

Notification

22nd April, 1993

GSR. 386 (E).– In exercise of the powers conferred by section 6 and 25 of the Environment (Protection) Act, 1986, the central government hereby makes the following rules further to amend the Environment (Protection) Rules, 1986, namely:–

1. a) these rules may be called the Environment (Protection) rules, 1993
   b) they shall come into force on the date of their publication in the official gazette.
2. In the Environment (Protection) Rules, 1986:–
   a) in rule 14
   i) for the words "audit report" wherever they occur, the word "statement" shall be substituted.
   ii) For the figures, letters and words "15th day of May" the words 30th day of September shall be substituted.
   b) In appendix "A" for Form V, the following form shall be substituted, namely

"FORM V"

(See rule 14)

- Environment statement for the financial year ending the 31st March.

PART A

- Name and address of the owner/occupier of the industry operation or process

Industry category primary (SIC) code Secondary (SIC Code)
Production capacity - units
Year of establishment
Date of the last environmental statement submitted

PART B

Water and Raw Material Consumption

Water consumption, m^3/d
Process :
Cooling :
Domestic :

Name of the Product Process Water Consumption per unit of Product Output

	During the Previous Financial Year	During the Current Financial Year
(1)		
(2)		
(3)		

Raw Material Consumption

Name of Raw Material	*Consumption of Raw Material per unit of product output*	
	During the Previous Financial Year	During the Current Financial Year

Industry may use codes if disclosing details of raw material would violate contractual obligation, otherwise all industry have to name the raw materials used.

PART C

Pollution discharged to environment/unit of output
(Parameter as specified in the consent issued)

Pollutants	Quantity of Pollutants Discharged (mass/day)	Concentrations of Pollutants in Discharges (mass/volume)	Percentage of Variation from Prescribed Standard With Reasons
(a) Water			
(b) Air			

PART D

HAZARDOUS WASTES

(As specified under Hazardous Wastes/Management and handling Rules, 1989)

Hazardous Waste

	Total Quantity (kg)	
	During the Previous Financial Year	*During the Current Financial Year*
(a) From process		
(b) From pollution Control facilities		

PART E

Solid Wastes

	Total Quantity	
	During the Previous Financial Year	*During the Current Financial Year*
(a) From process		
(b) From pollution control facility		
(c) (i) Quantity recycled or reutilised with the unit		
(ii) Sold		
(iii) Disposed		

PART F

Please specify the characterisation (in term of composition and quantum) have hazardous as well as solid wastes and indicate disposal practice adopted practice adopted for both these categories of wastes.

PART G

Impact of the pollution abatement measures taken on conservation of natural resources and on the cost production.

PART H

Additional measures/investment proposal for environmental protection including abatement of pollution, prevention of pollution.

PART I

Any other particulars for improving the quality of the environment.

ANNEXURE-II

QUESTIONNAIRE FOR ENVIRONMENTAL AUDIT

PART I

1. Name of the industry & location :
   (Enclose a map showing surrounding land uses upto 500m)
2. Date of commencement of production:
3. Type of industry:
   (i) LARGE/MEDIUM/ SMALL
   (based on capital investment)
   (Small: <Rs 50 lacs; Medium:Rs 50 lacs - Rs 5 crore; Large: > Rs 50 crore)
   (ii) MANUFACTURING/ FORMULATING /MFG. & FORMULATING
4. Products manufactures

Name of the product.	*Installed Capacity, (t/a)*	*Avg.production (t/a)*	*No. of days of prod. Per year*	*Remarks*

Do you have separate process lines for each product? Yes/No
If no, state which products are manufactured in the same process line?

5. Raw materials used, product wise (kg/t of product):

6(a) Details of manufacturing process alongwith process flow chart showing various unit operations and material balances. Indicate point sources emission and fugitive emission (Enclose separately)

(b) Give details of the State-of-art technology for the products manufactured:

(c) Do you have plans for expansion? Yes/No
If yes, give details

(d) Do you have plans for the modernisation of process technology? If yes, provide details. Yes/No

PART II

7. Water requirement:

(a) Industrial (product wise):

Name of the product	Process operation where water used	Quantity used, kl	
		Per day	Per t of Product

(b) Domestic : kld;
(c) Others : kld;
(d) **Total** : kld
(Also incorporate all the above figures in PROCESS FLOW CHARTS)

8) Wastewater generated :

(a) Industrial:

S.No.	Operation where waste generated	Quantity used, kl	
		Per day	Per t of Product

(b) Domestic : kld
(c) Boiler : kld
(d) Cooling tower : kld
(e) Others (specify) : kld
(f) Total : kld

(Also incorporate the above figures in process flow chart)

9(a) Characteristics of wastewater, stream-wise :

Product	*Source of wastewater Generation*	*Characateristics*

b) Characteristics of combined wastewater

Parameter	*Before Treatment*	*After treatment*	*Standards Prescribed*

10. Effluent treatment system provided :

Describe the treatment system provided to individual streams and combined effluent. Enclose drawings of the effluent treatment system showing sizes of the individual units. Also provide analysts reports showing the performance of the individual treatment units. (Enclose separately)

11. Treated effluent disposal :

(a) DOMESTIC:

(b) INDUSTRIAL:

(Specify the place of disposal and mode of discharge using sprinklers, marine outfall, etc.)

12. Storm water drainage system:

- Is it separate from industrial/domestic drainage system?
- Describe the method of storm water collection, treatment and final disposal.

13)(a) Solid waste :

Source	*Composition*	*Quantity*	*Mode of Collection*	*Mode of Disposal*

(b) Hazardous waste :

Source	*Composition*	*Quantity*	*Method of Collection*	*Method of Disposal*

PART III

14. Details of emission generated :

(a) Process emission (product wise) :

S.No.	Source *	Type of pollution	Source Strength (mg/Nm^3)	Std prescribed	Gas Flow Nm^3/ hr	Control Equip-Ment **	Conc. In Stack/ Vent (mg/Nm^3)	Stack Height (m) above GL root level

* Indicate the operation in the reactor, incinerator, flare, furnace, etc.

** Attach drawings of air pollution control systems alongwith design data

*** Based on actual monitoring

(b) Emissions from combustion of fuel viz. Coal, fire wood, furnace oil, etc.

Fuel used		Purpose	Type of control equipment provided	Gas flow in Stack mg/Nm^3	Conc. Of SPM/ SO_2 in Stack/ Vent Mg/Nm^3	Chimney Height (m) above GL/Rool Level
Type	Qty. (kg/hr. Day)					

c) Information regarding principal air contaminants (Fugitive emissions):

Activity	Air contaminant	Typical toxic concentration (Range)	Possible sources of emission
Eg. Polyvinyl Chloride	Vinyl chloride	20 mg/cum	Leaks in Pressurised system

d) Are emissions monitoring provisions made in the stacks/ vents? Yes/no

15. Ambient air quality in the factory premises :

Parameter	Concentration (Annual) μg/m^3 (Based on actual monitoring)		
	Minimum	Maximum	Average
SPM SO^2 CO NOX Acid mist HC VOC (non-methane) Others, specify			

PART IV

16. Details of implant pollution control measures:
17. Details of waste (wastewater, gaseous emissions, solid waste) minimisation techniques used and Recycling/reuse of waste adopted :
18. Whether the standards and other conditions laid down by the State Pollution Control Board under the Water Act, Air Act, and Environmental (Protection) Act are complied with?

	Has valid consent/ authorisation	*Complying with stds. and other conditions*
The Water Act, 1974 :	Yes/No	Yes/No
The Air Act, 1981 :	Yes/No	Yes/No
The Environment Protection Act, 1986 :	Yes/No	Yes/No

If no, specify reasons thereof
(Also enclose copies of consent orders)

19. Details of monitoring facilities available with the industry for emissions, wastewater, solid waste and the receiving environment

Parameters	*Instrument Used*	*Analytical Technique*

20. Details of organisational set up for environmental management :

S.No.	*Name of the person*	*Designation*	*Qualification*	*Duties*

21. Cost of pollution control :

(a) Annual sales turnover of the industry : Rs. ....

(b) Fixed cost of pollution control system (PCS) : Rs...............
(Mention details of components included)

(c) *Operating cost of PCS : Rs.............*

(d) Annual burden of PCS (Annualised capital cost of PCS + Operating cost) : Rs................

22. Are there any public complaints against pollution from? your industry? Yes/No
If yes, give details.

23. Details of green belt/plantation:

24. Problems faced by the industry in pollution control:

25. Name of the contact person with designation and telephone No :

Signature ....
Name : ....
Designation:..

Date :
Place:

ANNEXURE-III

ENVIRONMENTALLY SAFE LAYOUT CODE FOR MANUFACTURING UNITS

1. An environmentally safe layout plan takes care of material loss, cost of collection disposal recycle and treatment which are parts of the process itself, and consequently of the layout arrangement.
2. This layout code postulates that environment protection is a factor for designing any equipment reaction vessel material transfer arrangement, storage tank and service support to operate the production system.
3. All places of storage of solid and liquid materials are to be diked without drains. Any spillage is to be wiped out and cannot be washed out.
4. Each vessel should have its own catchpit to collect spills.
5. Each pump must be mounted on its catchpit; a suction line of the pump should be connected to empty the pit, periodically or regularly or continuously.
6. As losses of materials take place during charging of the reaction vessels, discharging of produce and dripping of outlet valves and as materials may be either solid or solid slurry or liquid, care needs to be exercised to prevent the losses if necessary by changing the charging/discharging and transfer devices.
7. In order to collect spills from a particular vessel before the spilled materials get a chance of contamination with spills from another nearby vessel the two vessels must be installed at sufficient distance so that inter-contamination cannot take place. The extra distance, ' non-contaminating distance' is to be provided for recycle of materials.
8. Flange joints should be avoided wherever avoidable.
9. Corrosion-prone areas and construction materials liable to atmospheric and process induced corrosion should be given special attention for finding better replacement material and stricter preventive maintenance frequency.
10. Exhaust ducts and fan outlets are sources of pollution, if the thrown out air is contaminated with pollutants . These may be treated before vented .Any vapour line should be

connected with either a recovery system or an absorption system.

11. The engineering code for the operation of pressurised systems and the established practice for preventive maintenance are consistent with the protection of the environment. These system are fitted with pressure release valves, and in many cases with rupturable discs. The present practice is to allow the released materials to the atmosphere. To be environmentally, safe, these lines ought to be connected to recovery/adsorption/ absorption arrangements .The rupturing of safety discs is accompanied with sudden release of high pressure; the design of the recovery arrangement of the released materials should be befitting the sudden emerging conditions of high temperatures/ pressure/ volumes.
12. New units will build floors with expanded metals, slotted angles, steel grills, steel grates prefabricated industrial floor gratings, and the like which will make floor washing redundant.
13. If the pant layout demands that vessels should be installed in upper floors, arrangements should be simultaneously made to spill avoidance /collection. Vulnerable points of leakage should be taken special care of. This is necessary not only for pollution control but also for the safety of plant personnel working in lower floors.
14. Storage tanks of raw materials for supply to the production vessel. Should be installed on a separate structure located just outside the main plant building with arrangement for holding spills and overflow. Level alarms should be installed where possible; where the same is not feasible because of the nature of the liquid, two overflow pipes at two different levels of the tank should be fitted .
15. Plant management should evolve its own code for washing equipment, where a particular equipment is used for the manufacture of different products. Dry scraping of equipment surface followed by moping with wet cloth should be carried out before hosing operation. This will reduce the quantity of contaminants and wastewater volume.
16. All channels be fitted with wastewater measuring devices half barrier for the separation floating immiscible liquid and in

built separation/sedimentation basins for withholding settleable particulate matters. This provision may be treated as compulsory for wastewater channels in the immediate vicinity of wastewater generating units.

17. All water usages that do not come in contact with chemicals, should have no opportunity to mix with process water. Uncontaminated water should have separated outfits from the plant and if recycle is not possible, should be drained out though separate channels, without any change of getting contaminated.
18. This proposed layout code recognises the solid waste generated in the process of manufacture must find a place within the factory premises. It will be stored on land /lagoon which will be lined with compatible geo-textile materials.
19. The detoxification operation is to be carried out outside the main production plant and provision has to be kept for the same.
20. Strom water drains should be segregated form process water drains the former may be used for the removal of cooling water and non-process water

    (Source; Minimal National standards-Pesticides manufacturing and formulation industry', COINDS/15/ 1985-86, Central Pollution Control Board, Delhi)

ANNEXURE IV

GUIDELINES TO MINIMISE MATERIAL LOSSES AND WASTES

HOW TO REDUCE RAW MATERIAL LOSSES ?

- Keep only an appropriate inventory of raw materials to ensure minimum material handling losses evaporation losses etc.
- Adopt mechanical handling of materials with proper monitoring facilities so as to dose only the predetermined quantities as per norms prescribed
- Plant layout should be properly made so as to minimise transfer distance of materials between storage and process or between unit operations
- There is a risk of cross contamination due to usage of same storage tanks for different materials depending on the batch product . Separate storages are to be provided
- Separate process lines for separate products or separate equipment for each unit operation can minimise losses due to residues left out in the equipment which are usually washed out
- Storage tanks should be provided with proper dip arrangements for exhausts /vents and insulation provided so as to reduce evaporation losses
- Enclosed and covered material storage areas keep them secured and reduced losses due to carry over by wind and rain.
- Enclosures should be made to collect spills and overflows of materials at the material transfer and sampling points. These, .if collected properly, can be recycled.
- Regular maintenance should be taken to check flange leaks, breaks/cracks, pump failures etc.
- Raw material purity should be ensured. Viscous raw materials lead to losses due to residues in Drums. Raw materials should be easy to handle Good house-keeping practices should be followed.
- Normal for performance of various process operations fixed so that the material usages are minimised and hence the material losses.

HOW TO REDUCE WATER USAGE AND WASTEWATER GENERATION?

- Quantities required for each operation should be determined and water usage regulated strictly. Reduced water usage reduces wastewater. Good house-keeping practices reduce water usage.
- Spills of material should be restricted to enclosures constructed for this purpose. The floor washings can then be minimised and at times totally avoided
- Wastewater may be stored and reused. The storage cost may be lower than waste treatment and disposal costs.
- Storm water drains should be kept separate and provisions should be made to collect only the rainfall of first few hours which carries contaminants. This can be subsequently treated and disposed.

HOW TO REDUCE EMISSIONS?

- The process operations where emissions arise should be providing with control equipment. Condensers can collect certain emissions, which can be entirely reused.
- The transfer of materials should be done through closed operation.
- The areas where fugitive emissions arise and can be avoided should be enclosed and the air-exhausted through induced draft and passed through control equipment before venting off. The enclosed area should be provided with at least three air replacements per minute.
- Evaporation losses from storage tanks should be checked by proper insulation and putting the vents in suitable dip columns.
- Loading and unloading of materials from tankers leads to huge quantities of emissions. The material transfers should be done through pipes/holes keeping the outlet of the tanker and the inlet of receiving tank covered. While loading the tanker, if the tanker inlet cannot be overrode, a hood can be provide over the inlet and emissions collected through a ducting system and further controlled.
- The scrubbing of gaseous emissions with a suitable chemical can yield a useful by-product. Recycle or recovery of useful by-products thus can avoid the discharges. The wastewater is usually treated up to secondary treatment level to conform to the required standards. By providing tertiary treatment by dual media filtration, chlorination, activated carbon filtration etc. wastewater can be reused for floor wash, gardening, toilets etc.

GLOSSARY

Biodiversity	–	A global term referred as ...all aspects of biological diversity especially including species richness, ecosystem complexity and genetic variation
Biosphere	–	The part of the earth that includes living organisms
CAD	–	Computer Aided Design
CFC's	–	Chlorofluorocarbons
Cerecla	–	Comprehensive environmental response, compensation and liabilities act (used in UK)
COSHH	–	Control of substance hazardous to health (UK)
CPCB	–	Central Pollution Control Board
CTREE	–	Council for training and Research in Ecology & Environment
DoE	–	Department of environment
EA	–	Environment Auditing, a review of environ-mental conditions and the environment impact of the activities of the particular activities of a particular company or institution, at a local level it is divided into two parts a) the external audit referred some time as State of the Environment and b) the internal audit also referred as review of policies and practices,
EIA	–	Environment impact assessment
EMAS	–	Eco-management and audit scheme, an EU initiative designed for private sector, now adopted for use by UK local authority which consist of a set of formalised procedures for addressing the environmental impacts of an organisation's policies and practices
EPA	–	Environment Protection Act
GIS	–	Geographical Information System
GNP	–	Gross National Product
IA	–	Internal Auditing
IFS	–	Indian Forest Service
LCA	–	Life Cycle Analysis, assessment of the environmental impact of a particular product or service
LGMB	–	Local Government Management Board, widely used by EU countries
MA	–	Management Audit
MoA	–	Ministry of Agriculture

MoE&F	–	Ministry of Environment and Forests, Government of India
NAEB	–	National Afforestation and Eco-Development Board
PIA	–	Policy Impact Assessment, Evaluation of the environmental impact of an agency's regulatory, policy and service activities
RIP	–	Review of Internal Practices, an evaluation of environmental impact of an agency's own operations and practices
SEA	–	Strategic Environmental Assessment
SoE	–	State of Environment, sometimes known as a external audit, it involves regular monitoring and review of the quality of the environment in the area concerned, and of implication for global sustainability
SPCB	–	State Pollution Control Board
VOC's	–	Volatile Organic Compounds
WHO	–	World Health Organisation

Council for Training and Research
in Ecology & Environment
C-18/19, Qutab Institutional Area,
Vishwa Dharmayatan Sansthan Campus,
New Delhi-110068, Tel.: 91-011-6867345

Dear Sir,

Please find enclosed herewith the questionnaire which has been compiled to get more information about the environment auditing process in corporate organisations and local authorities as well.

As you may be well aware about the environment auditing, which has been made mandatory for the polluting industries and it is also desired from the local authorities such as the State Governments, Municipal Authorities, Development Authorities and other statutory Local Authorities. In the last few years, the environment auditing process has been initiated with a small beginning. And, still there is a lot of research to be carried out in order to improve the audit models and procedures.

The questionnaire is prepared with a purpose to get the knowledge and experiences of Industries and Local bodies, while dealing with the matter of auditing.

In the annexed format, **Part one** is structured for those industrial and local authorities who has already carried out an audit or who are in the process of carrying out the environmental audit and **Part two** is meant for those who have not undertaken an environmental audit as yet.

So far established components of an audit, it is little understood by the industry/company/or, local authority and there is a confusion over the environmental audit process. We have included here the idea of UK where environment audit is defined as Internal Audit, State of Environment report and the mix of the two.

Looking forward to receive the filled up questionnaire at your earliest, which will certainly help us in formulating the auditing procedure in a greater way.

Thank you,

Prof. A.K. Shrivastava
CE: CTREE
Or
Mr. Vikram Singh, RD : CTREE
Phone : 6868269

APPENDICES

APPENDIX-I

Part 1 : Local Authorities/Industries/Companies who have undertaken, or are in the process of under-taking, an environ-mental audit

Identification

i. Name of Local Authority / Company :

ii. Name of Respondent :

iii. Department Section/Post held :

iv. Date :

The purpose of the audit

1. Bearing in mind the definition of an audit outlined in the covering letter, with which type of audit has the local authority been involved ?
   1. State of the Environment Report (SoE) ☐
   2. Internal audit (IA) ☐
   3. Full/Comprehensive Audit (SoE/IA) ☐
   4. Other - please-specify ☐
2. What stage has the audit process reached ?
   1. Completed ☐
   2. In progress ☐
   3. Being planned ☐
3. What were the stated objectives of the audit ?
   1. Heightening awareness about the condition and needs of the environment. ☐

2. Enabling the authority to raise the profile of environmental issues. ☐
3. Helping the authority/company to play an enhanced role in resolving environmental problems. ☐
4. Increasing public access to environmental information. ☐
5. Widening public participation in environmental decision making. ☐
6. Providing a baseline of environmental data in order to monitor progress and change. ☐
7. Identifying gaps in environmental data. ☐
8. Encouraging and co-ordinating co-operation between the many agencies - public, private, and voluntary, which have an environmental role. ☐
9. Enabling the authority/company to assess the environmental impact of their policies and practices. ☐
10. Producing an action plan, charter, or strategy for the environment. ☐
11. Providing a resource for educational use. ☐
12. Engendering the concept that environmental quality and economic well-being are complimentary ☐
13. Providing information to assist the authority/company in their enabling/advisory role ☐
14. Improving the image of the authority as an environmental role model. ☐
15. Other - please specify. ☐

Who set the objectives of the audit ?

1. Chief executive or his/her office. ☐
2. Environmental Forum ☐
3. Policy Committee ☐
4. Environment Committee/Sub-committee ☐
5. Officers working group ☐

6. Other–please specify ☐
5. None of the above ☐

5. Before the audit was undertaken, did the authority/company have any of the following ?

1. A general environmental policy. ☐
2. An environmental or 'Green' charter. ☐
3. A strategy for the environment. ☐
4. Any other specific environmental policies - please specify ☐
5. None of the above ☐

6. Have you referred to any of the following documents while preparing the audit ?

1. Environmental Charter for Local Government/Corporate sector ☐
2. Environmental Practice in Local Government/Corporate Sector ☐
3. Managing the Environment - local authority/company in action. ☐
4. The Environmental Role of Local Government/ Company ☐
5. Other local authority/industries audits ☐
6. Department of Environment (MOE & F) ☐
7. Other - please specify ☐
8. None of the above ☐

The Management of the Audit

7. Which of the following procedures were employed while carrying out the audit :

1. Policies and practice of the authority/ company listed, with general comments. ☐ ☐
2. Collection and analysis of secondary data about the local environment. ☐ ☐
3. Primary surveys on the state of the environment. ☐ ☐

4. Data for each environmental subject area assessed against pre-determined indicators or standards; ☐ ☐

5. Use of a comparative matrix assessing the impact of each of the authorities policies and practices against each environmental subject area. ☐ ☐

6. Other - please specify ☐ ☐

8a. Is there a unit or section within the local authority/company which has dealt specifically with the audit ?

1. Yes (refer Question 8b) ☐
2. No (refer Question 9) ☐

8b. Briefly explain the nature and role of this unit

9. Which members committee deals with environmental initiatives ?

1. Policy Committee ☐
2. Environment Committee/Sub Committee ☐
3. Planning Committee ☐
4. Environmental Health Committee ☐
5. Other - please specify ☐

10. Was an inter-departmental officers working group established with responsibility for the audit ?

1. Yes ☐
2. No ☐

11. Have any of the following staff appointments been made or planned in connection with the audit ?

1. Environmental Co-ordinator ☐
2. Environmental Scientist ☐
3. Researcher ☐
4. Database Manager ☐
5. Other - please specify ☐
6. No new appointments ☐

12. What mechanism has been used to link officers and councillors ?

    1. Officers - members co-ordinating group ☐
    2. Environmental co-ordinator ☐
    3. Representatives from officers on the principal environment committee ☐
    4. Informal contacts ☐
    5. Other - please specify ☐

13. Were outside consultants employed at any stage of the audit ?

    1. To help set the aims and scope of the audit. ☐
    2. To undertake the audit ☐
    3. To assist internal staff in carrying-out the audit ☐
    4. To review an internal audit ☐
    5. Other - please specify ☐

14a. What was the budget for the audit ?

Initial budget — Rs.

For implementation of actions — Rs.

14b. What are the main budgetary elements ?

15. How many person hours has the audit taken to complete ?

    1. Specify number of hours : ☐
    2. Estimate of number of hours ☐
    3. No estimate possible ☐
    4. Not applicable ☐

Consultation Process

16. Were the public consulted as part of the audit ?
    1. Opinions sought to set terms of reference. ☐
    2. Opinions sought on perception of environmental quality ☐
    3. Opinions sought to verify audit findings ☐
    4. Opinions sought on options for action ☐
    5. Other - please specify ☐

    6. No survey of public opinion carried out ☐

17. Was an environmental forum established to represent the views of other agencies - public, private and voluntary ?
    1. Yes ☐
    2. Already in existence ☐
    3. No ☐

18. How was the commitment and involvement of members of staff gained ?
    1. Staff bullet in/news letter ☐
    2. Special information sheets ☐
    3. Departmental committees ☐
    4. Departmental 'Green' representatives ☐
    5. Informally ☐
    6. Other - please specify ☐

    7. None of the above

19. How important, in the context of the audit, is the influence a local authority/company can have on other groups policies and practices through its role as regulator.

    Very important ☐ ☐ ☐ ☐ ☐ Not important at all

Scope of the Audit

20. What subject areas/issues have been covered in the audit ?

Environmental Category	Areas of assessment		
(Tick boxes as appropriate)	*State of the environment*	*Impact of local authorities own practices*	*Impact of local authorities policies on the environment*
Geology & Soils			
Air Quality			
Water Quality			
Waster & recycling			
Noise			
Energy			
Urban land use			
Rural land use			
Landscape & townscape			
Wildlife			
Open space			
Transport			
Food & agriculture			
Economy & work			
Purchasing policy			
Estate management			
Consumer advice & protection			
Environmental education			
Investment policy			
Other - please specify			

21. What sources of information have been used for the audit ?

    1. Primary surveying ☐
    2. Data already existing within the local authority/company ☐
    3. Central government ☐
    4. Local voluntary organisations and community groups ☐
    5. Environmental NGOs ☐
    6. Environmental Protection Agencies (CPCB, SPCB, DoE (MoE&F etc.) ☐
    7. Private industry ☐
    8. Academic institutions ☐
    9. Other local/company authorities ☐
    10. Local authority/corporate sector ☐
    11. Other–please specify ☐

22. Was a database used to compile the data ?

    1. Geographic Information System (GIS) ☐
    2. Other computer system ☐
    3. Manual database ☐
    4. None ☐

23. How was the local authority's/company's position on each issue assessed ?

    1. Predominantly Indian limits or values ☐ ☐
    2. Predominantly limits or values ☐ ☐
    3. Best practice, nationally ☐ ☐
    4. Where possible best practice, internationally ☐ ☐
    5. Targets set in an environmental strategy ☐ ☐

6. Other - please specify ☐ ☐

7. No attempt at assessment. ☐ ☐

Achievements of the Audit

24. How successful has the audit been at co-ordinating the activities of different departments within the local authority ?

Very successful ☐☐☐☐☐ Very unsuccessful

25a. How closely has the final audit report matched the stated objectives ?

Very closely ☐☐☐☐☐ Not at all

25b. Can you identify some areas of success and failure ?

25c. Can you analyse SWOT.

26. How effective was the chosen methodology found to be in meeting the objectives set for the audit ?

Very effective Note at all effective

27. How would you assess the contribution the audit has made to raising environmental awareness within the local authority/ company's surroundings and in the area as a whole ?

(Tick as appropriate)	Within local Authority	Within area
1. Significant increase in awareness	☐	☐
2. Slight increase in awareness	☐	☐
3. No change	☐	☐
4. No assessment possible/Don't know	☐	☐

29. Were reports/summary prepared/published.

	Public documents	Items for sale	Internal
1. State of the Environment summary	☐	☐	☐
2. Internal audit (IA) report	☐	☐	☐
3. Internal audit summary	☐	☐	☐

4. Joint SoE/IA report ☐ ☐ ☐
5. Joint SoE/IA summary ☐ ☐ ☐
6. Strategy for the environment ☐ ☐ ☐
7. Environmental action plan ☐ ☐ ☐
8. Information bulletins/newsletters ☐ ☐ ☐
9. Other - please specify ☐ ☐ ☐

10. No publications ☐ ☐ ☐

29a. Were recommendations for action a part of the audit ?

1. Yes ☐
2. No ☐
3. They will form a later stage of the auditing process ☐

29b. Were these recommendations :

	Yes	No	Some
1. Costed	☐	☐	☐
2. Prioritised	☐	☐	☐
3. Delegated to specific staff	☐	☐	☐
The Next Step	☐	☐	☐

30. What is the next stage in your auditing process ?

1. Produce an IA report ☐
2. Produce a SoE report ☐
3. Produce a SoE/IA report ☐
4. Formulate an action plan ☐
5. Implement recommended actions ☐
6. Public consultation on findings ☐
7. Internal debate of findings ☐
8. Other - please specify ☐

31. Have procedures have been established to monitor performance on issues raised in the audit ?

1. Yes ☐
2. No ☐
3. Partially ☐

32. Can you list the 3 most significant actions which have taken place as a result of the audit ?

 1.

 2.

 3.

33. How often will reviews of the audit take place ? (tick as appropriate)

	Full Review	Partial Review
1. Every year	☐	☐
2. Every 2 years	☐	☐
3. Every 3 years	☐	☐
4. Every 4 years	☐	☐
5. Every 5 years	☐	☐
6. More than 5 year intervals	☐	☐
7. No review planned	☐	☐

Thank you very much for helping with this survey. Please have a quick look back through the questionnaire to ensure that all questions are answered in full. A prompt return of the questionnaire, in the envelope provided, would be much appreciated.

Finally, I would be grateful to receive any documents your authority have produced relating to environmental auditing, or details of those available for purchase.

Thank you for assistance.

CTREE
Council for Training & Research in Ecology & Environment
C-18-19, Qutab Institutional Area,
New Delhi-16

APPENDIX-II

Part 2 : Local Authorities/Industries/Companies who have not yet undertaken an environmental audit

Identification

i. Name of Local Authority :

ii. Name of Respondent :

iii. Department/Section/Post held :

iv. Date :

A possible future audit

1a. Do you intend to carry out an audit in the future ?

1. Yes (refer Question 2) ☐
2. Undecided (refer Question 2) ☐
3. No (refer 1b) ☐

1b. If you have decided not to carry out an audit, briefly explain why not.

2. Bearing in mind the definition of an audit outlined in the covering letter what type of audit would be most appropriate for your authority ?

1. State of the Environment Report (SoE) ☐
2. Internal Audit (IA) ☐
3. Full/Comprehensive Audit (SoE/IA) ☐
4. Other - please specify ☐
5. Don't know ☐

3. Which of the following objectives do you think an audit should fulfill ?

1. Heightening awareness about the condition and needs of the environment. ☐
2. Enabling the authority to raise the profile of environmental issues. ☐
3. The environmental issues the authority to play an enhanced role in resolving. ☐
4. Increasing public access to environmental information. ☐
5. Widening public participation in environmental decision making. ☐
6. Providing a baseline of environmental data in order to monitor progress and change ☐
7. Identifying gaps in environmental data. ☐
8. Encouraging and co-ordinating co-operation between the many agencies-public, private, and voluntary, which have an environmental role. ☐
9. Enabling the authority to assess the environmental impact of their policies and practices. ☐
10. Producing an action plan, charter, or strategy for the environment. ☐
11. Providing a resource for educational use. ☐
12. Engendering the concept that environmental quality and economic well being are complimentary. ☐
13. Providing information to assist the authority in their enabling/advisory role. ☐
14. Improving the image of the authority as an environmental role-model. ☐
15. Other–please specify. ☐

4. Do you think audits should be conducted using ?

   1. In house team (existing staff). ☐
   2. In house team (new specialist staff) ☐
   3. External consultants ☐
   4. A combination of the above ☐

5. Have any of the following documents been brought to your attention ?

   1. Environmental Charter for Local Government/Corporate sector. ☐
   2. Environmental Practice in Local Government/Corporate Sector. ☐
   3. 'Mananging the Environment - local authorities/company in action. ☐
   4. The Environmental Role of Local Government/ Company ☐
   5. Other local authority/company audits. ☐
   6. Department of Environment. ☐
   7. Environmental Auditing in Local Government/Company ☐
   8. None of the above. ☐

   The Local Authority and the Environment/The Industry and the environment ☐

6a. At present, does any evaluation of environmental policies and practices take place ? (financial, quality, achievements etc.)

   1. Yes (go to Question 6b) ☐
   2. Partially (go to Question 6b) ☐
   3. No (go to Question 7) ☐

7. Does the local authority/company have any of the following?

   1. A general environment policy ☐
   2. A strategy for the environment ☐

3. A green/environment charter ☐
4. None of the above ☐

8. Which parts of the authority are responsible for environmental initiatives ? (i.e. the development of new policies and programmes, monitoring strategies etc.) Mark with an asterix the lead agent(s) for environmental initiatives

1. Policy Committee ☐
2. Environment Committee/Sub-committee ☐
3. Planning Committee ☐
4. Environmental Health Committee ☐
5. Environmental Co-ordinator ☐
6. Planning Department ☐
7. Environmental Health Department ☐
8. Inter-departmental Officers Working Group ☐
9. Officers - Members Co-ordinating Group ☐
10. Other - please specify ☐

9a. Is there a separate budget for environmental initiatives ?

1. Yes (refer Question 9b) ☐
2. No (refer Question 10) ☐

9b. How much is this budget, annually ?

Rs.

10a. Has a public opinion survey on environmental issues been carried out in the area ?

1. Yes (refer Question 10b) ☐
2. No (refer Question 11) ☐

10b. How have the results of this survey been utilised ?

11. Should the local authority/company be aware of environmental impacts at the following scales ?

1. Globally ☐

2. Nationally ☐
3. Regionally ☐
4. Locally ☐

12. Briefly describe the three most recent environmental initiatives undertaken by the authority ?

    1.

    2.

    3.

13. Does an environmental forum exist to represent the views of other agencies :-

    1. Yes ☐
    2. Being considered ☐
    3. No ☐

Thank you very much for helping with this survey. Please have a quick look back through the questionnaire to ensure that all questions are answered in full. A prompt return of the questionnaire, in the envelope provided, would be much appreciated.

Finally, I would be grateful to receive any documents, your authority have produced relating to environmental policies and plans, or details of those available for purchase.

Thank you for your assistance.

A.K. Shrivastava
CE : CTREE

PARAMETERS

Minars for Pesticides industry

Compulsory parameters	Limiting concentration
Temperature	
Shall not exceed 5C above the receiving water	temperature
PH	6.5 to 8.5
Oil and grease	10 mg/l
Bio- chemical Oxygen Demand	30 mg/l
Total suspended solids	100 mg/l
Bio-assay test	90% survival after 96 hrs with fish at 100% effluent

Optional Parameters (in mg/l)	Limiting concentration
a) Specific pesticides	
● Benzene hexachloride	10
● Carbaryl	10
● DDT	10
● Endosulfan	10
● Dimethoate	450
● Fenitrothion	10
● Malathion	10
● Phorate	10
● Methylparathion	10
● Phenthoate	10
● Pyrethrums	10
● Copper oxychloride	9,600
● Copper sulphate	50
● Ziram	,000
● Sulphur	30
● 2,4D	300
● Paraquat	23.000
● Propanil	7,.300
● Nitrofen	780
● Phosalone	80

(b) Heavy Metals	
• Copper	1.0
• Manganese	1.0
• Zinc	1.0
• Nickel	1.0
• Mercury	0.01
• Tin	0.1
(c) Organics	
• Phenol and phenolic compounds as C_6H_5OH	1.0
(d) Inorganics	
• Arsenic as As	0.2
• Cyanide as CN	0.2
• Nitrate as NO_3	50
• Phosphate as P	5

Notes

1. Limits : at the end of the treatment plant before any dilution
2. Bio-assay test : carried out with available species of fish in receiving water
3. State Boards prescribes total dissolved solids (TDS), sulphate and chloride depending on the uses of recipient water body. Industries are advised to analyze pesticides in wastewater by advanced analytical method such as GLC.
4. State Board prescribes COD limit correlated with BOD limit.
5. Pesticides are known to have metabolites and isomers. If they are found in significant concentration, standards prescribed for these in optional list by State Board.

Other Parametres

Biochemical Oxygen Demand 20°C, 5 days mg/l
Chemical Oxygen Demand mg/l
Dissolved Oxygen mg/l
Total residual chlorine as Cl_2 mg/l
Oil and Grease mg/l etc.

Besides the other parametres, such as pollutants specific to the industry should also be analysed and reported.

1. Soap and Detergents
   Surfactants
2. Petroleum Refinery Sulphite
   Phenolic Compounds
3. Iron and Steel
   Manganese
   Zinc
   Lead
   Tin
   Copper
   Nickel
   Iron
   Phenolic Compounds
   Cyanide
4. Non-ferrous metal
   Aluminium
   Cadmium
   Selenium
   Zinc
5. Rubber
   Lead
   Zinc
6. Glass
   Phosphorous
7. Photographic Material
   Silver
   Cyanide
8. Electroplating
   Lead
   Cadmium
   Copper
   Nickel

Zinc
Silver -
Gold
Total Metal
Cyanide

9. Caustic Soda
   Mercury
10. Man Made Fibre
    Zinc
11. Cotton Textile and Woollen
    Sulphide
    Phenolic Compounds
12. Nitrogenous Fertilizer
    Vanadium
    Arsenic
    Cyanide
13. Dye and Dye Intermediate
    Mercury
    Copper
    Zinc
    Nickel
    Cadmium
    Phenolic Compounds
14. Pesticide
    Pesticide Chemicals
    Arsenic
    Copper
    Manganese
    Zinc
    Mercury
    Tin
    Phenolic Compounds
    Cyanide
15. Uranium/Radium
    Uranium
    $Radium^{228}$ (Total)
    $Radium^{228}$ (Dissolved)
    Radioactivity

NATIONAL COUNCIL FOR REGISTRATION OF ENVIRONMENTAL AUDITORS
(under the aegis of Council for Training & Research in Ecology and Environment)

C-18/19, Qutab Institutional Area, New Delhi-110 068

Dear Sir,

We take this opportunity to introduce the NCERA, as a body that registers Environment Auditors and Impact Assessors to specific level of competence. The NCERA has opened National Register for registration of the following experts and consultants as grouped hereunder :

1. Environment Auditors or Environment Impact Assessors, and Environmental Experts, having five years working experience in the related field.
2. Trained Environment Auditors or Environmental Impact Assessors, who wish to undertake assignments for Environmental Impact Assessment from the Corporate Sector organisations as per Indian EIA/EA Regulations.
3. Executives working with the Departments of Environment, Community Development, Agriculture Development, Horticulture, Floriculture, Geology, Geomorphology, Hydrology, Social Science, Economics, Business Administration, Research and Development, Law, Welfare, Earth Studies, Education, Micro-Biology, Botany, Zoology, Disaster Management, NGOs, Public or Private Sector-who are keen to learn advance Environment Assessment or Environment Auditing procedures.
4. Environmental Activists, Lawyers, Builders, Engineers, Managers, Decision Makers, Planners, Consultants, Social Activities and all those, who are concerned with the EIA regulation's and EA procedure.

The NCERA has a provision to provide environmental auditors to the companies-institutes-industries either Governmental or Private, NGOs or all those concerns, who have a wish to prepare periodical environmental audit or environmental assessment/evaluation work on their environmental conditions or requirements, with the help of external consultants or to get their selected sponsored or nominated executives to be trained.

Keeping in view to make the registered members well informed about the subject, a periodical journal on EA/EIA/ESM is being produced. A National/International Seminar/workshop/conference will also be organized time to time, specially for the members as grouped and pointed in Sl. No. 2, 3, & 4 respectively, with the help of Sl. No. 1 and other prominent environmentalists. Participants shall be awarded certificates for their participation in the Seminar/workshop/conference.

Since, India has enacted the National Environmental Policy, Environmental Impact Assessment (EIA) and Periodical Environmental Auditing (EA) has become an importance feature of the environmental management in the entire sector who are concerned with EIA/EA regulations. The sectors concerned with the EIA/EA regulations are :

(i) Corporate Sector (Industrial & Commercial house)/ (ii) Public Sector (Industrial, Commercial and Engineering houses)/ (iii) NGO's dealing with Eco-Environmental matters/ (iv) Govt. organisations dealing with ecology & environmental issues.

Keeping in mind to provide necessary assistance to the Indian Companies, Indian Institute of Environmental Auditors, has also been formed under the banner of NCREA, which undertakes environmental auditing and assessment work on request of the Govt. Organisations, NGO's, Corporate bodies, Companies, Industries and Local Authorities. The IIEA has made a provision to enlist the organisation in the manner as stated above, with a view to provide assistance in the matters of environmental audit, environmental impact assessment and environmental safety management.

We would appreciate your willingness to register yourself with us as environmental auditors/assessors-individually through the NCREA and for institutional membership registration be ensured through IIEA.

With kind regards,

(Registrar)

NCREA (CTREE)

---

* Membership form : annexed.

NATIONAL COUNCIL FOR REGISTRATION OF ENVIRONMENTAL AUDITORS

(under the aegis of Council for Training & Research in Ecology and Environment)

C-18/19, Qutab Institutional Area, New Delhi-110 068

1. **Group One**
   Name :
   Address :
   Experience :
   Qualification :
   Completed and Undertaken Jobs :

2. **Group Two**
   Name :
   Address :
   Experience :
   Qualification :
   Area of Specialisation :

3. **Group Three**
   Name :
   Address :
   Experience :
   Qualification :
   Organisation where Associated :

4. **Group Four**
   Name :
   Address :
   Experience :
   Qualification :
   Area of Service :

5. **Group Five**
   Name :
   Address :
   Experience :
   Qualification :
   Organisation and their Concern with Environment :

---

* With other relevant materials/publications/achievements/interests in support of your concern with the environment. (attach documents).

INDEX

A

Accessibility 77
Accommodation 151
Acidification 101
Aesthetic audit 94
Aesthetic impact 80
Agriculture 49
Aims of
Environment auditing 12,
Local authority 60
Air 259
Air pollution 107
Air pollution 64
Air quality 112
Air quality 99
Air-water 25
Aluminium smelter 274
Aniline 237
Animal Species 51
Asbestos 139
Attitudes to recycling 123
Audit element framework 28
Audit team 248
Auditing 20
Auditing strategy 165

B

Bio-accumulation 110
Biodegradable 125
Biodiversity 39, 45
Blue Angel 135
Budgetary elements 202
Budgeting 190
Building 75
Built environment 40

C

Car sharing 84
Carbon dioxide 138
Caustic chlorine 239
Chlorofluorocarbons 134, 139
Choices of vehicles 81
Chronic bronchitis 237
Climate change 101
Climate change 55
Commercial buildings 76
Commercial environment 234
Commercial waste 122
Communication 111
Community 214
Community involvement 210
Community involvement 23
Compulsory competitive tendering 130
Condensing boiler 71
Conservation 93
Conservation 56
Consultants 206
Consultants 22
Consultative committees 145
Consumerism 132
Consumers 125
Contaminated land 102
Controlling pollution 241
Coordination 152
Corporate commitment 196
Cost implications 132
CPCB 278
Customer 130
Cycling 87

D

Decision making 189
Deforestation 181
Degree of sustainability 37
Delegation 190
Design 76
Development
  Control 54, 61,
  Planning 53, 61
Devin country council 73
Dioxins 101
Discrete element 188
Disposal of waste 117
Diverse decision 32
Documentation sources 185

E

Eco consumer guides 133
Eco green 132
Eco Management 9
Ecolabels 135
Eco-Management System 205
Economical element 21
Ecosystems 97
Education 57
Effluent treatment plant 275
Elements of Environmental Stock 36
Energy 138
Energy management 67
Energy planning 64
Environment 98
Environment monitoring 83
Environmental auditing 225, 233
Environmental Auditing 7, 47
Environmental awareness 147
Environmental awareness 154
Environmental capacity 38
Environmental coordinator 204
Environmental forum 216
Environmental health 261
Environmental information 157
Environmental Management System 227
Environmental pollution 180
Environmental pollution 242
Environmental purchasing objective 130

F

Fertiliser 274
Field observations 251
Fragile landscapes 42
Fuel burning 64

G

Gandhi Mahatma 3
Garden waste 122
General approach environmental auditing 266
Gentrification 95
Geographic Information System 185
Geographical database 25
Global ecology 35
Global pollution 80
Global sustainability 80
Green belt 213

H

Hazardous to health 140
Hazardous waste 123
Hazardous waste management 119
Heavy metals 100
Housing audits 75
Housing stock 68
Human environmental quality 36

I

Implementation 132
Implementation budget 218
Incineration 118, 122
Industrial pollution 262
Industry 240
Information 59
Infrastructure 36, 39
Infrastructure capacity 41

In-house 201
In-house auditing 20
Initial budget 217
Internal auditing 53
Internally focused 166

L

Land Management 53, 56, 61
Land pollution 109, 113
Landfill 117
Laudably pragmatic approach 178
Lead 141
Life cycle analysis 131

M

Macro issues 13
Maintenance 81
Management 57
Management audit 11
Material balance 249
Matrix approach 27
Matrix system 26
Mechanisms Public involvement 148
Methodology 272
Microorganisms 252
Model management system 195
Monitoring 191
Monitoring 250
Monitoring system 65

N

Native species 57
Natural conservations 45
Natural resources 25
Natural resources 35, 41
Need of Environment Audit 12
Need of Environmental information 264
Network analysis 29
Noise 102
Noise pollution 108, 113

O

Ornithology 47
Ownership 207
Ozone 139

P

Park 85
Participation 177
Peat 57
Philosophy of Environmental audit 263
Plastic bags 116
Plastic rubbish sacks 116
Policy Assessment 18
Policy content 129
Policy impact assessment 10
Polluting industries 277
Pollution 236
Pollution 79
Pollution control audits 104
Pollution control system 271
Pollution implications 13
Pollution industries 243
Pollution policy issues 99
Post-audit activities 268
Preliminary information 247
Principle elements 10
Priorities 46
Prioritisation 189
Prioritizing the tasks 172
Public consultation 212
Public participation 159
Purchase 126
Purchasing process 128
Radiator system 72
Realistic explanation 176

R

Recommendations 16
Recycling 120
Recycling energy 181

Recycling system 115
Refuse collection 116, 121
Rehabilitation 70
Reliability 184
Renewable resources 129
Resilience 5
Resources 215
Resources 256
Resources 62
Robustness 5
Role of coordination 226
Role of Local authority 47
Rural community 144

S

Scooping 179
Scope 149
Scope of Natural environment 48
Single window 151
Solid waste 273
Spheres waste 8
Staff training 152
Status of issue 179
Steering group 198
Stoichiometric 252
Sustainability 123
Sustainability 43
Sustainable community 214
Sustainable development 174
Sustainable development 227
Sustainable development 89
Sustainable Development Defination 3
Synthesis of data 251

T

Temperature 107
Theory of unsustainability 4
Thermostatic radiator 71
Thorny economic issues 182
Timspan 188
Tools for auditing 25
Toxic metal 139
Traffic capacity 88
Transferability 38
Transport 88
Transport sector 79
Travel 85
Travel to work 82
Trigger 74
Turf roofs 73
Type of audit 166

U

Urban natural habitats 51
Urban park-lands 51

V

Vehicles 82
Vehicular emissions 78
Volatile organic compounds 100

W

Walking 87
Waste audit 255
Waste awareness 122
Waste flow 250
Waste treatment evaluation 252
Wastewater 268
Water 103
Water management 256
Water pollution 108, 113
Weak sustainability 37
Weather data 107
Wildlife habitats 42
Wood lands 50

Z

Zeus 4